Pesticides and Nitrogen Cycle

Volume I

Editors

Rup Lal, Ph.D.
Lecturer
Department of Zoology
Sri Venkateswara College
New Delhi, India

Sukanya Lal, Ph.D.
Scientist
Department of Zoology
Sri Venkateswara College
New Delhi, India

CRC Press, Inc.
Boca Raton, Florida

Library of Congress Cataloging-in-Publication Data

Pesticides and nitrogen cycle.

  Bibliography: p.
  Includes index.
  1. Pesticides—Environmental aspects—Collected
works. 2. Nitrogens cycle—Collected works.
I. Rup Lal, 1953-   . II. Sukanya Lal.
QH545.P4P4813   1987   589.9′504133   87-7985
ISBN 0-8493-4350-X (set)
ISBN 0-8493-4351-8 (v. 1)
ISBN 0-8493-4352-6 (v. 2)
ISBN 0-8493-4353-4 (v. 3)

International Standard Book Number 0-8493-4350-X (set)
International Standard Book Number 0-8493-4351-8 (v. 1)
International Standard Book Number 0-8493-4352-6 (v. 2)
International Standard Book Number 0-8493-4353-4 (v. 3)

Library of Congress Card Number 87-7985
Printed in the United States

Dedicated to

Amma, C. L. Kaushal, and Kamla.

PREFACE

The indispensable role of microorganisms in sustaining soil fertility through nitrogen cycle is well known. The major events of nitrogen cycle are ammonification, nitrification, denitrification, and nitrogen fixation. Any possible side effect of pesticide application on nitrogen transformations may have far-reaching consequences on the maintenance of soil fertility and crop production. As reduction in current extensive use of pesticides is unlikely, it is imperative to obtain adequate information on the effects of pesticides on microbes of nitrogen cycle.

The literature on the effects of pesticides on nitrogen cycle, though enormous, is widely distributed in different journals and books. An attempt has been made to collect some of this information and make it available in a single book. The microbes involved in nitrogen cycle, their habitats, ecological significance, and role of environmental factors in controlling their population and activities are described in the first chapter. The broad aspects of biochemistry and physiology of the nitrogen cycle have been briefly reviewed and updated also in that chapter. They illustrate the value of exploiting combined ecological and biochemical knowledge to study the problem of effects of pesticides in depth. The technique of detecting, measuring, and assessing the side effects of pesticides on nitrogen cycle are described in the second chapter. The subsequent chapters present exclusively the data on the effects of pesticides on nitrogen transformations.

One of the principal benefits of growing a leguminous crop is the symbiotic fixation of nitrogen. Rising costs of nitrogenous fertilizers together with the possible environmental problem make biological nitrogen fixation of further increasing agricultural importance. Legumes are also grown as a green manure in rotation involving cereals. Current practices employed for legume production also include inoculation of seeds with rhizobia and their treatment with pesticides to protect them from pathogens. Thus, it is important to consider any possible effects of pesticide on *Rhizobium*, host plants, and *Rhizobium*-legume associations. These aspects have been thoroughly reviewed for the first time in a chapter on *Rhizobium*-legume associations.

We thank the contributing authors for their interest in the project. We gratefully acknowledge the encouragement and support given by Dr. V. Krishnamoorthy during the preparation of the book. Thanks are also due to Ashok Kumar for secretarial help and M. L. Vergheese and A. S. V. N. Rao for typing the manuscript.

THE EDITORS

Rup Lal, Ph.D., has been in the Department of Zoology, Sri Venkateswara College, University of Delhi since 1979. Born on September 27, 1953 in Kanoh, Himachal Pradesh, India, he obtained his B.Sc. from D.A.V. College, Jullundur in 1973, He also obtained an M.Sc. from Kurukshetra University, and his Ph.D. degree in 1980 from the University of Delhi. Since then, he has been actively engaged in research and teaching. He is the coordinator of a project, "College Science Improvement Programme" (COSIP), from the University Grants Commission.

Dr. Rup Lal has been the recipient of many research grants from the Government of India. Currently he is the principal investigator of a project entitled "Effects of Pesticides on *Rhizobium*-Legume Associations" from the Department of Environment, Government of India. He is also associated with a project, "Survey of Pesticide Residues in Food Commodities", from World Health Organization, through the Ministry of Health, Government of India.

Dr. Rup Lal has been recognized by the University of Delhi as a guiding supervisor for Ph.D. students in Zoology, being the first lecturer in a college under the University to get that honor. His major research interests relate to the interaction of pesticide with the *Rhizobium*-legume associations and the use of microbes for detoxification of pesticides.

Sukanya Lal, Ph.D., is a research associate in the Department of Zoology, Sri Venkateswara College, New Delhi. Born on October 28, 1957 in Hamirpur, India, she graduated from Government College, Hamirpur in 1975. She received her M.Sc. and Ph.D. degrees, respectively, in zoology and pesticide microbiology from the University of Delhi, India. She is associated with a research project, "Effects of Pesticides on *Rhizobium*-Legume Associations", from the Department of Environment and Forests, Government of India. She has authored or coauthored approximately 15 research papers in the area of pesticide microbiology. Her current interests include accumulation, metabolism, detoxification, and effects of pesticides on microorganisms.

CONTRIBUTORS

P. S. Dhanaraj, Ph.D.
Scientist
Department of Zoology
Sri Venkateswara College
New Delhi, India

Surendar Kumar, M.Sc.
Scientist
Department of Zoology
Sri Venkateswara College
New Delhi, India

Rup Lal, Ph.D.
Lecturer
Department of Zoology
Sri Venkateswara College
New Delhi, India

Sukanya Lal, Ph.D.
Scientist
Department of Zoology
Sri Venkateswara College
New Delhi, India

V. V. S. Narayana Rao, Ph.D.
Lecturer
Department of Zoology
Sri Venkateswara College
New Delhi, India

Ramesh C. Ray, Ph.D.
Lecturer
Department of Botany
Uendrapara College
Orissa, India

N. Sethunathan, Ph.D.
Head
Department of Soil Science and
  Microbiology
Central Rice Research Institute
Cuttack, India

S. Shivaji, Ph.D.
Scientist
Center for Cellular and Molecular
  Biology
Hyderabad, India

TABLE OF CONTENTS

Volume I

Volume II

Volume III

Chapter 1

NATURE AND HABITATS OF MICROBES OF NITROGEN CYCLE

Rup Lal

TABLE OF CONTENTS

I. INTRODUCTION

The biological availability of nitrogen, phosphorus, and potassium is of considerable importance to soil fertility and crop production. Of all these plant nutrients, nitrogen stands out as the most susceptible to microbial transformations. This element is the key building block of protein molecule upon which all life is based, and it is thus an indispensable component of the protoplasm of plants, animals, and microorganisms.

Nitrogen undergoes a number of transformations and the functional pattern of these pathways and pool of nitrogen is commonly referred to as the nitrogen cycle (Figure 1). The elemental nitrogen of the atmosphere is the primary source or pool of nitrogen. The second pool of nitrogen is the biomass of organisms which after death, breakdown, and mineralization of the organic biomass releases ammonia.

The ammonium nitrogen is taken up by soil biomass and plants leading to its immobilization. In addition, this inorganic nitrogen is passed through the nitrification process which constitutes a crucial point in the cycling of nitrogen. The nitrate-nitrogen is also liable to heterotrophic immobilization thereby constituting an enlargement of the above-mentioned mineralization-immobilization cycle to plant uptake and denitrification. In the case of den-

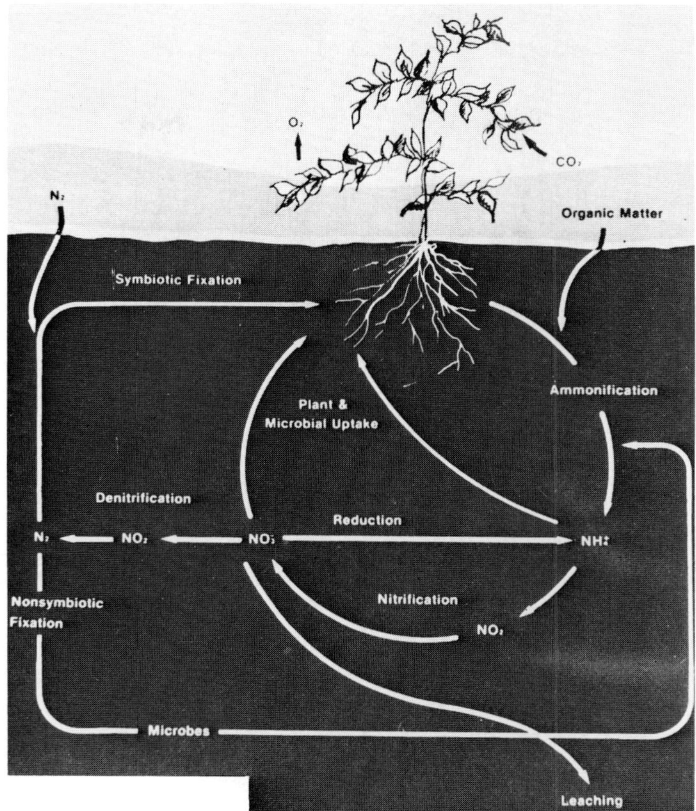

FIGURE 1. Nitrogen cycle

itrification, nitrate-nitrogen is reduced to the elemental form and joins the atmospheric nitrogen pool. This way the primary nitrogen cycle of nitrogen fixation, mineralization, nitrification, and denitrification, is closed.

The nitrogen cycle becomes enlarged when a plant biomass is consumed by animals (and man), in which case it is animal biomass and excreta that becomes liable to mineralization and humus formation. Finally, it must be emphasized that the pool of humus nitrogen is not completely resistant to biological attack. It is liable to slow mineralization resulting in a recycling of humus nitrogen.

The entire nitrogen cycle described above is mainly governed by microorganisms and is composed of several individual transformations. In nitrogen mineralization, part of a large reservoir of organic complexes in the soil is decomposed and converted to inorganic ammonium and nitrate. Microbial mineralization results in the degradation of proteins, polypeptides, amino acids, nucleic acids, and other organic compounds. In contrast, the conversion of the complex to the simple substances is nitrogen immobilization or assimilation. Microbiological immobilization leads to the biosynthesis of the complex molecules of microbial protoplasm from ammonium and nitrate.

Inorganic nitrogen in the form of ammonium ions is converted to nitrite and nitrate by microbes. Nitrogen, once in the nitrate form, may be lost from the soil in several ways. Because of its solubility in soil solution, nitrate readily moves downward out of the zone of root penetration. Nitrate and ammonium should also be removed to satisfy the nutrient demand of the plant cover. The greatest biological leak in the otherwise closed cycle in soil is through microbial denitrification whereby nitrogen is removed entirely from the realm of ready accessibility because the end-product of denitrification, N_2 is unavailable to most organisms.

Any leaks in the cycle deplete the soil's nitrogen reserve which eventually could have drastic effects on man's agricultural economy. As the leakage to the atmosphere is omnipresent, there exists a reverse process to maintain the balance. Atmospheric N_2 is acted upon by certain microorganisms, sometimes in symbiosis with higher plants, which can use nitrogen source for growth. This process, called nitrogen fixation, results in the accumulation of new organic compounds in the cells of the responsible organisms. The N_2 thus fixed re-enters general circulation when the newly formed cells are in turn mineralized.

Our main interest in nitrogen cycle centers around the effects of pesticides on microbes involved in nitrogen cycle. The nitrogen cycle is influenced by the use of pesticides which enter the soil system directly or indirectly. For the reliable evaluation of the effects of these pesticides on the nitrogen cycle, extended knowledge of the microbes involved is required.

II. NITROGEN MINERALIZATION

Nitrogen mineralization is the conversion of nitrogen of the organic compounds to inorganic forms like ammonium ions (NH_4^+), nitrate ions (NO_3^-), and others. Breakdown of organic compounds like proteins, nucleic acids, and related materials mainly releases amino acids. The ultimate fate of amino acids in the soil may be their utilization as nutrients by microorganisms or their degradation by microbial attack. The microbial degradation of amino acids occurs by a variety of pathways. The major pathway is deamination which is accomplished by the removal of the amino group from the amino acid. Although several variations of deamination reactions are exhibited by microorganisms, one of the end-products is always ammonia. An example of specific deamination reaction follows.

$$CH_3\,CHNH_2COOH + \tfrac{1}{2}\,O_2 \xrightarrow[\text{deamination}]{\text{Alanine}} CH_3\,CO\,COOH + NH_3$$
Alanine   Pyruvic acid   Ammonia

This reaction is classified as an oxidative deamination and the production of ammonia is referred to as ammonification. Many microorganisms can deaminate amino acids and the fate of ammonia produced varies, depending upon conditions in the soil. The ammonia produced either accumulates or is utilized by plants and microorganisms, and under favorable conditions oxidized to nitrates.

A. Microbiology

Population estimates on different soils reveal that 10^5 to 10^7 organisms per gram of soil are ammonifiers and the number of these organisms is dependent on the amount of nitrogen compounds available as substrates in the culture medium. A simple amino acid has a vast group of microorganisms acting upon it while the nitrogen of the chitin is freed only by certain select genera. In addition, in natural environments the breakdown of proteins and other nitrogenous substances is the result of the metabolism of a multitude of microbial strains each of which occupies a large or small niche in the pathway of degradation. Microorganisms of diverse nature have been reported to liberate ammonium from organic nitrogen compounds. The rate of decomposition and the compound utilized depend upon the nature of microorganisms.

The nitrogen mineralization occurs outside the cell by extracellular enzymes. These enzymes, particularly proteolytic enzymes, are synthesized by microorganisms which are used for decomposition of complex proteins. Aerobic bacteria, fungi, and actinomycetes as well as certain facultative and strict anaerobes constitute the proteolytic population of microbes. In addition to the production of ammonium and nitrate, CO_2, sulfate, and water, many other intermediates are formed during proteolytic action. These intermediates disappear quickly as they are quite unstable. The anaerobic transformations also lead to the formation of ammonia as the final product. In addition amines, CO_2, organic acids, indole, mercaptans, and hydrogen sulfide are also produced during the reaction.

The predominant species which decompose pure proteins readily are *Pseudomonas*, *Bacillus*, *Clostridium*, *Serratia*, and *Micrococcus*. In addition many fungi readily decompose proteins, amino acids, and other nitrogenous compounds with the liberation of considerable quantities of ammonium. Fungi, however, produce less quantities of ammonium than bacteria. These fungi also liberate proteolytic enzymes for the decomposition of proteins. Fungi dominate in acid soils while bacteria dominate in neutral and alkaline environments.

B. Environmental Influences

Physical and chemical characteristics of the habitat such as moisture, pH, aeration, temperature, and the inorganic nutrient affect considerably the rate of ammonification. In poorly drained land, the unsuitable water relationship depresses microbial metabolism, thereby lowering the rate of ammonification. Moderate and high moisture level generally favors the activity of aerobes and anaerobes involved in organic matter decomposition. A 50 to 75% water-holding capacity of soil is optimum for ammonification. The process of ammonification is also rapid where there is less oxygen, while flooding of the soil increases the quantity of ammonium because of the rapid degradation of organic molecules. Anaerobic conditions also favor the retention of ammonium which can be converted to nitrate only in aerated environments. The higher quantities of ammonium under particular conditions do not reflect a stimulation in ammonification because with further reduction in the conversion of ammonia to NO_2^- or NO_3^- will lead to the accumulation of NH_4^+-N. In many flooded soils and shallow lake sediments, conditions exit under which both nitrification and denitrification can proceed at the same time thus reducing the levels of NH_4^+-N.

III. NITRIFICATION

Nitrification is commonly defined as the biological oxidation of ammonium to nitrate with nitrite as an intermediate. The definition is rather limited because some heterotrophic microorganisms can produce nitrite or nitrate from reduced forms of nitrogen besides ammonium.[1-7] For this reason, Alexander et al.[8] proposed that nitrification be defined as the conversion of reduced nitrogenous compounds either organic or inorganic to others having nitrogen in more oxidized state. The importance of this process is that oxidized forms can participate in denitrification, permitting potential loss of N from the system.

A. Microbiology

The microbial nature of nitrification was first realized over 100 years ago when it was shown that the appearance of NO_3^- in soils and sewage was inhibited by antiseptics. Shortly after this it was established that there were two distinct groups of obligatory aerobic bacteria involved, both capable of obtaining energy at the expense of different N compounds. These bacteria are now grouped in the Nitrobacteriaceae family. Table 1 summarizes the currently recognized species and their habitats. With the exception of a few isolates of *Nitrobacter winogradskyi*, all are obligate chemolithotrophs. Bacteria with generic prefix "nitroso" carry out the oxidation of NH_4^+ where as the members of the second group with the generic prefix "nitro" oxidize NO_2^- to NO_3^-.

Much of our basic knowledge about nitrification in natural systems was provided by Winogradsky's classical work in the early 1890s leading to the recognition and isolation of

Table 1
NITRIFYING BACTERIA AND THEIR HABITATS

Substrate	Species	Habitat
NH_4^+ oxidized to NO_2^-	*Nitrosomonas europaea*	Soil, fresh water
	Nitrosospira briensis	Soil
	Nitrosolobus multiformis	Soil
	Nitrosovibrio tenuis	Soil
	Nitrosococcus nitrosus	Soil
	Nitrosococcus oceanus	Marine
	Nitrosococcus mobilis	Marine
NO_2^- oxidized to NO_3^-	*Nitrobacter winogradskyi*	Soil, fresh water
	Nitrospina gracilis	Marine
	Nitrococcus mobilis	Marine

a group of chemoautotrophic bacteria that oxidize ammonium to nitrate (the genus *Nitrosomonas*) and another group that oxidize nitrite to nitrate (the genus *Nitrobacter*). The information on nitrification in natural ecosystem and the mechanisms of nitrification has already been discussed in detail in several reviews.[6,7,9-18]

Generally, the enumeration and isolation of ammonium oxidizers is carried out in liquid inorganic media by most probable number (MPN) techniques but they can also be recognized on solid chalk containing media as colonies by clear zones. On the other hand NO_2^- oxidizers are more difficult to culture on a solid media, and are almost exclusively isolated by MPN techniques in a liquid media. *Nitrosomonas europeae* and *Nitrobacter winogradskyi* are the types most commonly isolated from soils, sewage, and the freshwater environment and these organisms are often considered to be the most important nitrifiers. They are also common in weathered rocks, and are often isolated in large numbers along with other nitrifiers from the surface of sedimentary rocks which may contain anything from 10 to 500 g NH_4^+/tonne.

Variable counts of nitrifiers in soil and sediments range from 10^3 to 10^5 organisms per gram, but larger numbers (10^7 to 10^8 organisms per gram) are generally described in high NH_4^+ environments such as activated sludge. Both groups of nitrifiers are roughly represented equally. It may however, be true that current estimates of the number of some species are mere underestimates, and are largely a reflection of cultural conditions used in enumeration procedures. An alternative explanation is that considerable nonlithotrophic nitrification also occurs in some environments. A number of soil microorganisms can also be organotrophic nitrifiers, including soil fungi such as *Aspergillus* spp., and certain soil bacteria like *Arthrobacter* spp. Both of these will produce significant amounts of NO_2^-, NO_3^-, and other oxidized N compounds when grown under conditions of high NH_4^+.

The bulk of global nitrification is usually attributed to the oxidation of inorganic nitrogenous compounds by chemolithotrophic bacteria such as *Nitrosomonas* and *Nitrobacter* spp.[6,7,18] Yet nitrification is by no means limited only to chemolithotrophs; numerous heterotrophic bacteria and fungi have the ability to oxidize a variety of nitrogenous compounds.[19] This is mainly due to the presence of different enzymes which can also oxidize ammonia.

The occurrence of substrates in natural environments well suited for heterotrophic (as opposed to autotrophic) nitrification is demonstrated by the accumulation of oximes by plants and animals,[20] and by the production of hydroxamic acids by a variety of microbes in pure culture.[21,22] Such compounds are found in soils under diverse climatic conditions,[23,24] while certain common soil bacteria such as *Arthrobacter* strains require them for growth.[6] The enzymes required for the breakdown of oximes can also be used by microorganisms for nitrification.

The most active heterotrophic nitrifier as yet reported is an *Alcaligenes* sp. isolated from

Table 2
DENITRIFYING
BACTERIA ACTIVE IN
NITRIFICATION

Alicaligenes faecalis
Chromobacterium violaceum
Flavobacterium sp.
Neisseria sicca
Paracoccus denitrificans
Pseudomonas aeruginosa
Pseudomonas aureofaciens
Pseudomonas denitrificans
Pseudomonas fluorescens 400
Pseudomonas fluorescens 401
Pseudomonas fluorescens 402
Pseudomonas stuzeri

soil.[25-28] This bacterium can oxidize pyruvic oxime but is inactive with other oximes.[28,29] It is also capable of assimilation of nitrate.[26] *Pseudomonas* has also been reported to act as a nitrifier.[30,31] Recently many denitrifying bacteria representing six genera have been found to be active in nitrification of pyruvic oxime or hydroxylamine (Table 2).[32] Thus, it would seem that the abilities to denitrify and nitrify heterotrophically are often found in the same organism in the environment. There may be advantages in having both abilities when O_2 partial pressure fluctuates in soils, as may occur after rain or temperature changes. Under aerobic conditions, heterotrophic nitrification could generate nitrite or nitrate which would then be available for denitrification after the onset of anaerobic conditions.

B. Environmental Influences

Important factors affecting nitrification include temperature and redox potential of soil. Nitrification is considered not to occur at redox values lower than $+200$ mV, and NO_2^- oxidation is more sensitive to changes in redox potential than NH_4^+ oxidation. NO_2^- also tends to accumulate at low temperature. Plant exudates suppress nitrification and there is some dispute as to whether nitrification is inhibited under climax vegetation.[6] The inhibition of nitrification under such conditions results in a massive decrease in NO_3^- levels in soils. However, it is possible that the low levels of NO_3^- under extensive plant cover is merely a consequence of plant demand. The large number of ammonium-oxidizers generally occur only at a pH greater than six. In most habitats, species of *Nitrosomonas* and *Nitrobacter* are found together. Populations of both the groups may be increased by the use of ammonium salts. In temperate climate nitrifiers are numerous during the warmth of the spring, rare during hot, dry summers, and in winter. Drying or freezing may decrease their abundance but cannot eliminate them completely.

Nitrite does not accumulate in soil and the dominant nitrogen-containing anion is nitrate. The existence and persistence of nitrite could have a marked effect upon crop production because of its toxicity to plants and microorganisms. On the other hand, nitrite may accumulate in certain circumstances in alkaline soils for example, nitrate production from (NH_4) SO_4 is suppressed as the rate of ammonium application is increased. This is not an inhibition of ammonium oxidation but rather the nitrite formed is not further oxidized to nitrate. In calcareous soils the accumulation of nitrite is proportional to the rate of ammonium addition and/or constant nitrogen fertilization. The effect however, increases with decreasing hydrogen ion concentrations. Even in the hydrolysis of proteinaceous matter or urea, the release of ammonium may lead to a significant though transitory suppression of nitrate formation. The nitrite accumulation is attributed to the marked sensitivity of the *Nitrobacter* group to

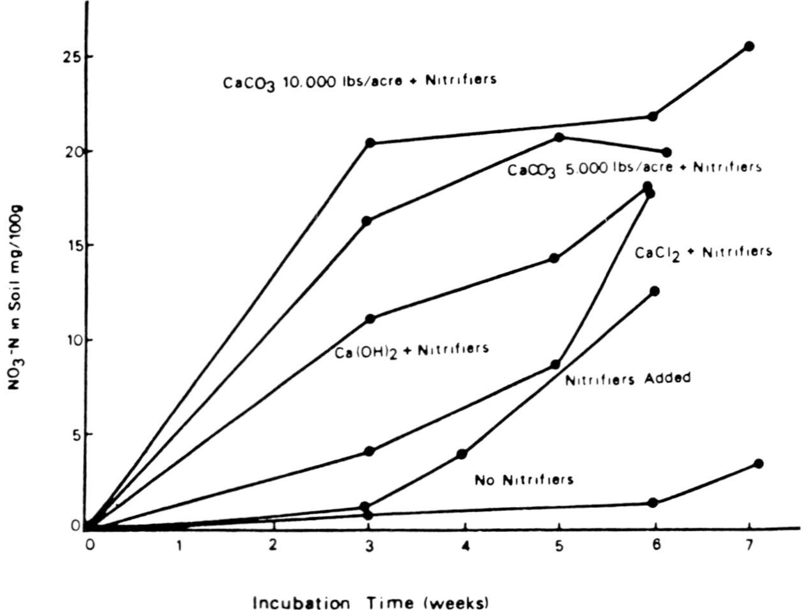

FIGURE 2.   Effect of nitrifiers and calcium sources upon nitrification in Bladen soil. (From Giddens, J., *Univ. Ga. Res. Bull.*, 327, 1-37, 1985. With permission.)

ammonium salts at alkaline reaction. At pH 9.5 as little as 1.4 ppm ammonium nitrogen suppresses the energy-yielding reaction of *Nitrobacter agilis* with no such marked influence on the ammonium oxidizers. This suggests that ammonium, the natural substrate for nitrification is a selective inhibitor in alkaline environments of the second step of nitrification by virtue of its toxicity to *Nitrobacter* spp. Nitrification has also been reported to be affected by land-disposed sewage sludge.[33-37] The decrease in nitrification in amended sludge treatment has been attributed to lowering pH as a consequence of metal salt addition in the sludge.[38] The application of high rates of sewage sludge containing high concentrations of metals may temporarily inhibit nitrification. Because of the low loss rate of most metals from soils, extended use of such sludges may seriously interfere with the nitrification. Nitrification may also be inhibited in some organic grassland soils by toxic substances from plant roots.[6] Giddens[6] reported that addition of nitrifiers to the soil increased nitrification and addition of different calcium sources further increased the rate of nitrification by neutralizing the toxic substances present in soil (Figure 2).

C. Biochemistry

Several mathematical models of nitrification have been proposed. Such models can be used to predict nitrifier biomass, growth rates, and concentrations of NH_4^+, NO_2^-, and NO_3^- under different conditions of NH_4^+ loading, and at different temperatures and oxygen concentrations. The biochemistry of nitrification in organotrophs and lithotrophs is probably very similar, but remains relatively obscure. During the six electron oxidation catalyzed by the "nitroso" group, hydroxylamine (NH_2OH) is almost certainly an intermediate and traces of nitric oxide (NO) and nitrous oxide (N_2O) are evolved. However, N_2O and NO do not seem to be utilized by NH_4^+ oxidizers, and are probably produced as a byproduct of some as yet undescribed intermediate. Accordingly the process may proceed in the following manner.

$$NH_4^+ \rightarrow NH_2OH \rightarrow HNO \rightarrow NO_2^- \rightarrow NO_3^-$$

Table 3
DENITRIFYING BACTERIA

Achromobacter	Gm⁻ rods
Acinetobacter	Gm⁻ rods
Agrobacterium	Gm⁻ rods
Alcaligenes	Gm⁻ rods
Azospirillum	Gm⁻ vibroids
Arthrobacter	Gm⁻ pleomorphic cells
Bacillus	GM⁺ rods
Chromobacterium	Gm⁻ rods
Corynebacterium	Gm⁺ rods
Flavobacterium	Gm⁻ rods
Hymphomicrobium	Gm⁻ swarmer cells
Kingella	Gm⁻ coccoid cells
Moraxella	Gm⁻ coccoid cells
Neisseria	Gm⁻ cocci
Paracoccus	Gm⁻ coccoid cells
Propionibacterium	Gm⁺ rods
Pseudomonas	Gm⁻ rods
Rhizobium	Gm⁻ rods
Rhodopseudomonas	Gm⁻ rods
Spirillum	Gm⁻ rods
Thiobaccillus	Gm⁻ rods
Vibrio	Gm⁻ bent rods

The process is poorly understood and the existence of HNO, which is very unstable has never been proved. The first enzyme involved in the oxidation of NH_4^+ to NH_2OH is an oxygenase which has a requirement for reducing power. N-serve operates by inhibition of this particular enzymatic step. A subsequent cytochrome linked oxidation of NH_4OH to NO_2^- occurs without detectable intermediates. The oxidation of NO_2^- is better understood and appears to be a cytochrome linked single-step hydrolytic oxidation.

IV. DENITRIFICATION

Many microorganisms have the capacity to reduce N oxides (NO_3^-, NO_2^-, NO, N_2O) under anaerobic conditions. In fact, denitrification is the reduction of these compounds followed by the generation of the gaseous products, N_2 and N_2O, which are lost from the system. This is also known as respiratory nitrate reduction which is quite different from assimilatory nitrate reduction in plants and microorganisms where NO_3 is reductively assimilated into tissues to fulfill N requirement. In the latter case the quantity of N being transferred is relatively small and subsequently immobilized, whereas in the former large amounts of N-oxides are reduced and released to the environment.

A. Microbiology

The ability to reduce N-oxides in a dissimilatory fashion (denitrification) is widespread among soil and aquatic microbes, including eukaryotes. Although not all microorganisms can carry out all potential reductive steps, more than 40% of soil and marine genera contain isolates capable of reducing NO_3^- to NO_2^- (Table 3) but further reductive steps leading to complete denitrification are restricted to bacteria of relatively few genera. Quantitatively members of chemoorganotrophic genera *Bacillus*, *Micrococcus*, and *Pseudomonas* are probably the most important in soils and *Pseudomonas*, *Aeromonas*, and *Vibro* in aquatic environment. It is not uncommon to isolate 10^6 denitrifiers per gram of soil or sediment. Most of the presently recognized denitrifiers are chemotrophic and heterotrophic. Only two or three function as chemolithotrophs. The sulfur oxidizers, *Thiobacillus denitrificans*,[39-41] utilize elemental sulfur and thiosulfate as the electron donors for denitrification. *Alcaligenes eutropha*,[42] *Paracoccus denitrificans*,[43,44] and several *Pseudomonas* species[45,46] can also grow lithotrophically as denitrifiers.

The majority of the denitrifiers are gram-negative rods, but several represent other morphological types. Notable among the structurally more complex are the *Hyphomicrobium* species distinctive in mode of reproduction and morphology.[47,48] In growth environments which they find acceptable, individual cells of these remarkable bacteria affix themselves at one end to a wet or submerged surface, then extend a tubular prosthesis from the free end. Buds formed at the end of each cell prosthesis are released in succession which mature to provide a population of rod shaped "swarmer" cells that can swim away.

Heterotrophic denitrifiers like *Pseudomonas* are preponderantly prototrophic aerobes with the capability for anaerobic respiration but not fermentation. Certain other (*Bacillus* and *Chromobacterium* species) live as aerobes, but each can live facultatively as either a fermenter or a denitrifier when oxygen is excluded.[49] The *Propionibacterium* species are the only nonoxygen-respiring, ordinarily fermentative bacteria known to be capable of switching to denitrifying but not aerobic respiration.[49,50] Several photolithotrophic bacteria are also known to participate in denitrification.[51-53]

The rhizosphere frequently contains large number of denitrifying bacteria.[54,55] Garcia[56] reported up to 18 million denitrifiers per gram of rice rhizosphere soil. The population of denitrifiers in the rhizosphere is markedly influenced by type of plant roots growing in the soil. Probably the first demonstration of a stimulatory effect of the plant roots on the denitrification activity in the rhizosphere was that of Woldendrop.[55] He also obtained evidence that the rates of addition of carbonaceous exudates and of consumption of oxygen could be sufficient to create the microenvironments in which denitrification could occur.[55] Garcia et al.[56,57] showed that the rice rhizosphere exhibited a significant role in the potential denitrification in rhizosphere as the denitrification rates were 14 times more than those of root free soil. Rice also shows a positive stimulatory effect on the rhizosphere denitrification even during the period of active photosynthesis.[58] This provides additional evidence that anaerobic microsites supplied with electron donors in the form of root exudates exist at some location, yet to be determined in the rice rhizosphere. It also appears that this rhizosphere phenomenon is quite sufficient to account for losses of 10 to 37% of fertilizer nitrogen from grassland soils[55] but it should be emphasized that losses through denitrification occur also in planted but unfertilized systems.[59]

N_2 has generally been considered to be the major product of denitrification and it is certainly true that in closed environments, the less fully reduced N products disappear with time and N_2 accumulates. The production of NO_2^-, NO, and N_2 by most denitrifiers suggests that the denitrification reduction pathway is as follows.

$$2NO_3^- \rightarrow 2NO_2^- \rightarrow 2NO \rightarrow N_2O \rightarrow N_2$$

The first enzyme in the sequence, nitrate reductase, is a membrane-bound iron and molybdenum-containing protein whose synthesis is repressed in the presence of O_2. Further reductions are carried out in a step-wise fashion by the other enzymes in the sequence, with cytochromes as electron donors, except in the case of nitrite reductase where a flavoprotein is involved.

Certain bacteria possess the entire pathway for carrying out complete denitrification and in a strict sense only these organisms are denitrifiers. However, the combined activities of organisms individually capable of catalyzing only one or a few steps of the pathway also bring about denitrification. Therefore, in an ecological sense, any organism that catalyzes any step of the pathway can contribute to denitrification, and hence is a partial denitrifier. Partial denitrification occurs for one of several physiological or genetic reasons.

1.   One intermediate of the pathway but not nitrate is available to the bacterium.
2.   Environmental conditions (O_2 concentration, pH, or concentration of an intermediate of the pathway) render one or more steps of the pathway nonfunctional.
3.   Owing to different rates of induction of the various nitrogen oxide reductases, only certain of these enzymes are present in cells at various times following the onset of induction.
4.   Certain bacteria are not genetically capable of synthesizing the complete array of nitrogen oxide reductases of the denitrification pathway. This last category is known to possess the following patterns of genetic variation in nitrification.
      a.   Organisms that are capable of reducing nitrate only to nitrite. These organisms lack nitrite, nitric oxide, and nitrous oxide reductases. This limited dissimilation of nitrate, sometimes termed nitrate respiration is widely distributed among bacteria. Organisms mediating nitrate respiration are probably the most numerous of the denitrifiers. Hall[60] has compiled a list of bacterial genera that, according to the 8th edition of Bargey's Manual, are capable of nitrate respiration (Table 3).
      b.   Organisms capable of reducing nitrate only to nitrous oxide. These organisms genetically lack nitrous oxide reductase. A number of bacteria clearly fit into this class. Some of them, e.g., *Asguaspirillium itersonii* and some strains of *Pseudomonas fluorescens* have been studied in laboratory culture for some time, making it unclear whether this pattern of partial denitrification represented among wild type strains is a consequence of the frequent mutational loss of nitrous oxide reductase or not. However, isolation by enrichment of strains that grow well on nitrous oxide as the terminal electron acceptor[61] implies that nitrous oxide producer either physiological or genetic occur with some abundance in nature.
      c.   Organisms capable of reducing nitrite but not nitrate to dinitrogen. These organisms are lacking nitrate reductase. Certain species of *Neisseria* fall into this class. This class presents distinct dilemmas in ecological studies because their inability to produce gas from nitrate causes them to be overlooked as denitrifiers by the usual microbiological tests even though their metabolic activities combined with those of the nitrate respirers bring about complete denitrification. The abundance of this class in nature is by no means clear, although it can only be speculated that usual microbiological procedures and enrichment techniques would not detect them so their number and ecological importance might be underestimated. The combined metabolic activities of these organisms and of the nitrate respirers result in complete denitrification.
      d.   Organisms capable of reducing nitrate to nitrite and nitric oxide to nitrous oxide. These organisms lack nitrite reductase and nitrous oxide reductase. This class is represented by a single report in the literature identifying *Bacillus licheniformis*[62] as having this capacity of partial denitrification. The pattern may be rare and hence of not much ecological significance.

Organisms capable of catalyzing complete denitrification (reduction of nitrate to dinitrogen) differ in their ability to utilize intermediates of the pathway, particularly nitrous oxide. Okerele[63] examined 77 strains of active complete denitrifiers isolated and characterized by Gamble et al.[64] — 64 strains were able to reduce exogenously supplied nitrous oxide, 13 strains could not utilize exogenous nitrous oxide, 1 strain could neither produce nor utilize exogenous N_2O, and 5 strains could not grow anaerobically when nitrous oxide was the only available electron acceptor. Some strains could reduce nitrous oxide but could not grow on it. These results suggest that if nitrous oxide is always an intermediate of the denitrification pathway, it is not always a freely diffusible intermediate and sometimes the reduction of nitrous oxide is not coupled to an electron transport chain capable of generation of ATP.

B. Environmental Influences

The magnitude and rate of denitrification is markedly affected by the environmental factors. The organisms which are responsible for denitrification have some physiological relationship to one another and changes in the habitat can appreciably stimulate or largely eliminate this population of denitrifiers. Chief environmental factors that influence denitrification are oxygen concentration,[65-69] alkalinity,[70] temperature,[71] organic matter,[72] dryness,[73] and soil activity.[70,74]

Denitrification of the nitrate in flooded soils is slower with low carbon contents than in soils rich in organic matter. In such soils denitrification is also enhanced by the addition of carbonaceous materials and the rate of denitrification is directly proportional to the addition of organic nutrients which are readily decomposed. Sugars or organic acids are more stimulatory than the less readily fermentable straw or grasses. Rotted plant and animal residues, however, have limited influence on denitrification as compared to the fresh materials.

Availability of oxygen is one of the major factors that determine the rate of denitrification. Aeration affects the transformation in two apparently contrasting ways. First, denitrification proceeds only when the O_2 supply is insufficient to satisfy the microbiological demand. Second, O_2 is necessary for the formation of nitrite and nitrate, which are essential for denitrification. Decrease in the partial pressure of oxygen enhances the denitrification of added nitrate. Soil aggregate size has a direct influence on gaseous exchange and the nitrogen losses decrease with increasing particle size because of the improved aeration. Thus, in general aeration enhances the rate of denitrification in soils receiving nitrate fertilizer. Similarly, denitrification will take place in paddy soils receiving ammonium salts, as the ammonium is oxidized in the upper zone of O_2 penetration, and the leaching downward of the resultant nitrate brings it into the underlying anaerobic zone. Nevertheless, the loss in the flooded soils is always greater for nitrate than for ammonium.

Over several decades, a few denitrifying *Pseudomonas* species were isolated in a routine survey of freshwater soils, salt marsh mud, and seawater. This was followed by the discovery of many denitrifiers from aquatic environments. Most of them require anaerobic conditions for denitrification, others are more tolerant to limited quantities of oxygen. One spiraled bacterium continues denitrifying at an oxygen tension as high as 153 mm. As might be anticipated nitrate accumulates transiently early in the denitrifying period in several of the isolates.[75-80]

Interest in eliminating nitrate from sewage prompted a search for the specific bacteria that released nitrogen at the expense of methanol — the simplest, most easily manageable organic source of electrons that can be used for this purpose. Surprisingly, anaerobic enrichment cultures supplied with methanol and nitrate yield pure cultures of a filamentous, budding *Hyphomicrobium* species that stoichiometrically releases nitrogen from nitrate or nitrite and carbon dioxide from methanol.[81] *Pseudomonas* species has also been reported to utilize benzoate and several other ring compounds as electron donors for anaerobic denitrifying growth.[82,83]

Denitrification is also related to the moisture contents in the soils. Denitrification of added nitrate is appreciable at high water levels and in localities having improper drainage. No losses of nitrogen occur at moisture levels below 60% of the water holding capacity regardless of the carbohydrate supply, nitrate concentration or pH. The effect of water is attributed to its role in governing the diffusion of O_2 to sites of microbiological activity.

The bacteria which bring about denitrification are sensitive to high hydrogen ion concentrations, and acid soils contain a sparse denitrifying population. The population becomes large only above a pH of approximately 5.5. Acidity governs not only the rate of denitrification but also the relative abundance of the various gases. The liberation of N_2O is pronounced in environments whose pH is low and frequently N_2O makes up more than half of the nitrogenous gases evolved from acid habitats. Likewise, NO only appears in significant quantities when pH is low. At neutral or slightly acid reaction, N_2O may be the first gaseous product, but it is reduced microbiologically so that N_2 tends to be the dominant product above pH 6. The difference in gas composition associated with pH may be largely due to the acid sensitivity of the enzyme system concerned with N_2O reduction. Nitrous oxide release in soil is further governed by nitrate concentration. The relative proportion of the gas is greatest at high nitrate levels.

Temperature also plays a significant role in deciding the rate of denitrification. The rate of denitrification is slow at lower temperature and increases with the increase in temperature. The optimum temperature for denitrification is 25°C. In certain cases the transformation is still high at elevated temperatures which may continue at temperatures as high as 60°C. This indicates that at higher temperatures thermophilic organisms participate in denitrification. Crop serves to reduce losses of nitrogen through denitrification and the existence of vegetable cover may markedly affect the magnitude of volatilization. This is due to the competition of vegetation with denitrifiers for the available nitrate thereby reducing the amount acted upon by microorganisms. For example, application of graded quantities of nitrate to a crop at intervals throughout the active growing season minimizes loss, whereas large-scale treatment of soil with nitrite before planting results in significant loss.[84,85] Even during active growth of a crop such as wheat, application of nitrate to soils poor in potassium provides the right conditions for extensive denitrification. This is because the plants that are physiologically slowed down by potassium starvation compete poorly with the soil bacteria for nitrate.[86] The flooding of soils in preparation for rice culture soon after spring thawing provides conditions for denitrification at the expense of the residual quantities of organic matter that have not been degraded during the colder months of winter. Flooding of the soil later in the spring, after the previous season's residue of organic material has been oxidatively degraded, diminishes loss of nitrogen.[87]

C. Biochemistry

As mentioned earlier denitrification process involves nitrate reduction and each step is controlled by individual nitrate reductase. The nitrate reductases are reported to be membrane bound. Initial events in the electron transport that results in nitrate reduction make use of pyridine nucleotides, flavins, and quinones, but when oxygen is in short supply, a branch from the oxygen-terminated electron-transport occurs at cytochrome b level.[88-95] Different c-type cytochromes are produced and used to transport electrons that reduce nitrate rather than oxygen. NAD, FAD, or FMN are revealed as cofactors for the electron transport leading up to the reduction of nitrate in several early studies. Quinones have also been shown as the components of the nitrate-reducing electron transport chain.

V. NITROGEN FIXATION

Fixation of atmospheric nitrogen is the major factor in nitrogen cycle. A part of the atmospheric nitrogen is artificially fixed by physical and chemical means. However, a major part of nitrogen is fixed microbiologically. With the development of new techniques such as Kjeldahl, ^{15}N and acetylene reduction methods to detect the ability of microbes to fix nitrogen, an increased number of species able to fix nitrogen have been identified. Such species and systems include free living system (bacteria and blue-green algae) and symbiotic system (rhizobia-legume, angiosperm-actinomycetes and blue-green algae associations). The agents of N_2-fixation must still be regarded as exclusively prokaryotic because the claim for fixation in eukaryote by Yamada and Sakaguchi[96] has not been confirmed. For the purpose of discussion this topic is mainly restricted to the organisms involved in nitrogen fixation and ecological factors that may influence the process of nitrogen fixation. Both areas are very broad and have been the subject of spectacular advances over the decade or two.[97-101]

Table 4
N₂-FIXING SYSTEMS

Free-living genera and groups

Aerobes	Microaerophilic	Facultative anaerobes	Anaerobes
Azotobacter	*Corynebacterium*	*Klebsiella*	*Clostridium*
Beijerinckia	*Azospirillum*	*Bacillus*	*Phototrophic*
Azomonas	Blue-green algae	*Citrobacter*	bacteria
Azotococcus	*Rhizobium*	*Erwinia*	*Desulfovibrio*
Methane oxidizers		*Enterobacter*	*Desulfotomaculum*
Heterocystous blue-green algae			
Thiobacillus			

Symbiotic associations

Nodule	Rhizosphere Temperate	Rhizosphere Tropical	Phyllosphere	Blue-green algae	Others
Legume + *Rhizobium*	*Bacillus* (nonspecific)	*Azospirillum* (nonspecific)	*Beijerinckia*	Lichens	Termites + Entrobacteria
Alder etc. + *Frankia*	*Klebsiella* (nonspecific)	*Beijerinckia* (nonspecific)	*Azotobacter*	*Azolla* + *Anabaena azolla*	
		Azotobacter paspali + *Paspalum notatum*		Cycads + *Anabaena* or *Nostoc*	Other animals + Enterobacteria

A. Free-Living Nitrogen-Fixing Bacteria

The free-living bacteria known to fix nitrogen can be classified as in Table 4. The list also contains organisms of soil, aquatic, phyllosphere and rhizosphere habitats, photosynthetic aerobic, anaerobic, and facultative species.

Among free-living bacteria family Azotobacteriaceae comprises four genera: *Azotobacter*, *Azomonas*, *Beijerinckia*, and *Derxia*. Genus *Azotobacter* has four species, e.g., *A. chroococcum*, *A. beijerinckii*, *A. vinelandii*, and *A. paspali*. Genus *Azomonas* has three species, e.g., *A. agilis*, *A. insigne*, and *A. macrocytogenes*. Genus *Beijerinckia* also has four species, e.g., *B. indica*, *B. mobilis*, *B. fluminesis*, and *B. derxii*. Genus *Derxia* has only one species, e.g., *D. gummosa*.

Azotobacter is one of the oldest and most common known nitrogen fixers. Its cells are often quite large. All azotobacters are strict aerobes and they apparently possess the highest respiratory rate. *Azotobacter chroococcum* appears to be the most widespread species in natural or alkaline soils.[102] *Azotobacter* is generally present in soil samples above pH 7.5, but below pH 6.5 they are rarely recorded. *Azotobacter vinelandii* is much less common in soil than *Azotobacter chroococcum*. The number of azotobacters is usually less than 10^4/g soil.[103] Higher numbers up to 10^7/g soil have been reported from irrigated alkaline clay soils which contained high levels of decaying plant residues.[104,105]

Estimation of the agricultural contribution of aerobic nitrogen-fixing organisms is difficult. The nitrogen fixed by the bacteria is usually incorporated into cells, which must be decomposed before inorganic nitrogen compounds are released. Thus, the nitrogen fixed is only indirectly available to the plant root systems. These bacteria utilize few nitrogenous compounds — N₂, ammonium, nitrate, urea, etc. Members of the genus are mesophilic and their optimum temperature is near 30°C.

The inoculant of *Azotobacter* is known as azotobacterin. Its cells are generally grown on agar in Roux bottles and harvested with water. The bacterial suspension is further diluted with water and used to spray seeds. This practice of spraying *Azotobacter* inoculant has been reported to increase plant yield significantly. The inoculant is generally prepared from *Azotobacter chroococcum.*

Other members of the family Azotobacteriaceae, notably *Azotobacter beijerinckii, Azomonas* spp., and *Beijerinckia* spp., are able to grow at lower pH values and are the dominant species in soil. The estimates of the abundance of beijerinckias in tropical soil have been recorded for soils associated with rice paddy[106] or sugar cane fields.[107] This number (20/g soil) recorded in nonrhizosphere soils is far less.

The cells of *Beijerinckia* are commonly smaller than azotobacter cells. These bacteria are generally confined to tropical soils. Members of the genus are also adapted to the phyllosphere or leaf surface of numerous plants growing in the tropics. *Derxia gummosa* occurs commonly in South American tropical soils with pH in the range of 4.5 to 6.5.

The anaerobic diazotrophs are primarily represented by species of *Clostridium* which are almost universally present in poorly drained agricultural soils, and to a lesser extent by *Desulfovibrio* sp. *Bacillus* and *Klebsiella* strains are the most commonly isolated facultative aerobes, but they fix nitrogen only under anaerobic conditions. They are widely distributed in a range of aquatic and terrestrial environments.[102] The population of anaerobic N_2-fixing bacteria is more dense in proximity to the plant root system. The more abundant and more completely investigated clostridia are of *Clostridium pasterianum.* In contrast with the *Azotobacter*, the *Clostridium* is found at sites of pH 5.0 and is still capable of growth at pH 9.0. Estimates of nitrogen-fixing activity of spore-bearing bacteria in anaerobic water-logged soils are less available than those for aerobic organisms in well-drained soils. Paul et al.[66] and Brouzes et al.[108] using carbohydrate amended soils in natural grasslands, reported that clostridia increased from 10^3 to 10^7/g soils and considerable amounts of nitrogen were fixed. Jensen[109] estimated that *Klebsiella pneumoniae* strains isolated from an aquatic environment fixed 4 to 5 mg N/g carbohydrate consumed.

1. Environmental Influences

The occurrence and functioning of free-living, N_2-fixing bacteria in natural environments (soil or water) is dependent on the number of growth requirements. These requirements include: the presence of available carbon compounds, the presence of adequate amounts of inorganic nutrients like calcium, magnesium, molybdenum, etc., an optimum pH for growth of the organisms, and a relatively low oxygen supply.

a. Presence of Available Carbon Compounds

The development of free-living, N_2-fixing bacteria in soil is favored by the presence of considerable amounts of available carbon compounds and very low amounts of combined nitrogen so that the C/N ratios are high. If the environments are well-provided with combined nitrogen, nonnitrogen-fixing microorganisms will readily develop and successfully compete with the nitrogen fixers for carbon compounds.[110] Examples of environments with large numbers of free-living N_2 fixers are given by Jensen,[109] Abd-el-Malek,[104] Rouquerol,[111] and Dobereiner et al.[112]

Jensen[109] treated nitrogen-poor, water-saturated soil with 1 to 2% of oats or wheat straw which had a high C/N ratio and reported the fixation of considerable amounts of N_2 by free-living soil bacteria of the *Azotobacter* type. The numbers of these organisms amounted to 10^6 to 10^7/g of soil. Similar high number of azotobacters have been counted in irrigated alkaline clay soils of the Nile Valley, containing large amounts of easily decomposable crop residue[104,105] and in rice soils of the Camargue in the Rhone delta.[111]

The occurrence of azotobacter in numbers amounting to 10^6/g of soil indicates that considerable amounts of nitrogen are fixed so that in this case the free-living N_2 fixers contribute substantially to the nitrogen economy of the soil. Normally, much lower numbers of azotobacters are counted (10^3 to 10^4/g of soil) when the pH is 7 or higher and the contribution of these organisms to soil fertility is insignificant.[108] Soils amended with large amounts of glucose and incubated under anaerobic conditions contained more than 10^8 *Clostridium* cells per gram which fixed considerable amounts of N_2.[108] Similar results were obtained by Paul et al.[66] with water-logged soil supplied with large amounts of straw.

A further example of stimulated N_2 fixation by specific free-living bacteria in soil resulting from an increased supply of particular carbonaceous compounds may be encountered in the vicinity of leaking natural gas mains. The most important component of this gas (methane) is readily oxidized by microorganisms of the Methylosinus type.[113] As some strains of this type are able to fix N_2 they will obviously accumulate nitrogen compounds selectively and will give rise to an increase in the nitrogen content of such environments.

Selective enrichment of certain N_2-fixing bacteria has been observed in the rhizosphere and in the phyllosphere of certain plants, particularly under tropical conditions. Beijerinckias have been found to accumulate in the rhizosphere of sugar cane[114] and in the phyllosphere of several tropical shrubs.[115,116] A particular rhizosphere organism is *Azotobacter paspali* which has been found by Dobereiner and co-workers to occur mainly in the rhizosphere of *Paspalum notatum*, a tropical grass growing in large areas of Brazil.[117] The relatively large amounts of N_2 fixed in this environment by *Azotobacter paspali* suggest that a particular association between *Azotobacter paspali* and *Paspalum notatum* may exist.[117]

Klebsiellas have been found by Evans and co-workers[118] to occur in large number in the rhizosphere of nodulated leguminous plants. Enterobacters were found in the rhizosphere of maize plants.[119] Bessems[120] isolated klebsiellas from the leaf surface and from the liquid inside leaf sheaths of maize and tropical grasses, particularly guatemala grass. He made a study of carbohydrate content and of the C/N ratio of the sheath liquid of these plants. Maize plants growing under tropical conditions excreted much greater amounts of carbohydrates than similar plants growing in temperate regions. Removal of the carbohydrate-containing solution from the guatemala grass by washing, stimulated the excretion of carbohydrates so that within a few hours considerably higher carbohydrate concentrations were attained than those existing before the washing started. In spite of these apparently highly favorable conditions for N_2-fixing bacteria, the amount of N_2 fixed was low, indicating that an essential factor for N_2 fixation was lacking.

b. Presence of Inorganic Nutrients

Inorganic nutrients like molybdenum, calcium, and iron are critical for nitrogen fixation in bacteria. Molybdenum is required for the metabolism of N_2, but microorganisms do not use nitrate unless molybdate is present although molybdenum requirement for nitrate utilization is less than for N_2 fixation. Growth on ammonium salts, however, proceeds rapidly in the absence of added molybdenum. In a like manner iron salts are implicated in the N_2 metabolism of *Azotobacter*, *Clostridium*, *Aerobacter*, and *Achromobacter*, but the specific requirement for N_2-metabolism is often difficult to establish because iron is required, to a lesser extent, for growth upon fixed compounds of nitrogen.

Although soils are poor in available inorganic nutrients like calcium, magnesium, molybdenum, and some others, very little is known about the effect of adding these nutrients on nitrogen fixation and growth of free-living bacteria in such soils. The adverse effect of a low soil pH on soil azotobacters may partly be due to calcium deficiency, as acid soils are very low in available calcium ions. Very low amounts of available molybdenum are found in iron stone-containing soils in Europe and Australia. Although symbiotic N_2 fixation of several leguminous plants growing on such soils suffers badly from Mo deficiency,[121] no data are available concerning free-living N_2 fixers.

c. Effect of pH

Most azotobacters require a neutral or alkaline reaction for growth. A survey of 264 Danish soils of widely different pH revealed that practically 100% of the soils above pH 7.5 contained azotobacters in numbers varying 10^2 and 10^4/g, whereas below pH 6.5 only a small fraction of the soils tested contained a few *Azotobacter*.[109] Although beijerinckias, when grown in pure culture, tolerate a wide range of pH values, they are found under natural conditions mainly in acid tropical soils.[122] Derxias grow in acid as well as in neutral soils.[117] (For more details on the effect of pH as ecological factor see Mulder and Brotonegoro.[123])

d. Oxygen Supply

Oxygen does not only adversely affect anaerobic and facultatively anaerobic N_2 fixers but to some extent it also interferes with N_2 fixation of the aerobic bacteria. This might indicate that azotobacters would prefer deeper layers of terrestrial and aquatic environments, where the oxygen supply may be expected to be lower. In this connection the experiments of Tschapek and Giambiagi[124] may be mentioned. These authors found that in nitrogen-free liquid media, *Azotobacter chroococcum* contained in tubes, developed a growth zone at a certain distance from the surface where the partial pressure of oxygen apparently was optimum. When the carbon compound content of the nutrient solution was higher, the growth zone moved upward, owing to increased oxygen demand. That a reduced oxygen supply may favorably affect N_2 fixation by rhizosphere organisms was shown by Dobereiner et al.[112] in experiments with sugar cane which contained beijerinckias as free-living N_2 fixers in the rhizosphere.

2. Biological Factors

Many associations exist between free-living diazotrophs and other organisms.[125] Only associations important in relation to the effects of pesticides on nitrogen fixation are described below.

a. Interaction Among Microorganisms

This type of interaction is difficult to define with precision because the intensity and nature of the interaction is itself often subtle and transient. A consensus of the literature, however, testifies to an enhancement in activity of free-living diazotrophs by microorganisms which do not fix nitrogen. For example, organisms capable of decomposing polysaccharides have been observed to stimulate nitrogen-fixation of *Azotobacter*, *Beijerinckia*, and *Clostridium* possibly by supplying carbon sources.

b. Phyllosphere

The phyllosphere association involving free-living diazotrophs and aerial parts of higher plants have long been overlooked in nitrogen fixation studies. Yet as early as in 1956 it was shown that the leaves of tropical plants generally supported extensive surface populations of microorganisms, among which aerobic diazotrophs were prominent.[126] Studies of tropical grasses show that these organisms exude significant concentration of carbohydrates, but little or no nitrogenous nutrients, thus providing a particularly suitable habitat for diazotrophs. Such associations have also been indicated in temperate crops and have been shown to have the potential for a significant contribution to the nitrogen economy of temperate zone woodlands. Ruinen[126] enumerated microorganisms of the phyllosphere in different parts of the world. Common among them are *Azotobacter*, *Beijerinckia*, *Derxia*, *Agrobacterium*, *Pseudomonas*, *Xanthomonas*, *Mycoblana rubra*, *Spirillum*, *Myconostoc*, *Flavobacterium*, *Aerobacter*, *Bacillus*, *Clostridium*, *Nocardia*, etc.

c. Rhizosphere

Hiltner (1904) first used the term rhizosphere to that part of the soil in which roots generally induce a proliferation of microorganisms. Since then many studies both, qualitative and quantitative, have been carried out and it has become evident that the metabolic activities of the microbes in the root region are of vital importance for the development of plant. Among these activities, nitrogen fixation is prominent and nitrogen-fixing bacteria are frequently reported to occur in the rhizosphere. This includes the aerobes *Azotobacter, Beijerinckia, Derxia,* and *Sprillum,* the facultative anaerobes *Bacillus, Enterobacter,* and *Klebsiella,* and the anaerobic *Clostridium.*[127,128] Rhizosphere contains the following important parts.

i. Mucigel

A gel-like layer or mucigel is a coating of the plant root system and had been described by many workers.[129-139] However, this layer is not observed in the rose or the chrysanthemum.[129] Mucigel extends from the soil particles to the epidermis or root cap cells. It is frequently reported in the root hair region either as a roughly continuous layer or in sterile roots.[138] Bacteria are frequently observed within the mucigel. Some of these bacteria, are surrounded by their own capsular polysaccharides.[130,131] In the mucigel, proliferating bacteria also produce their own slime material. This "bacterial mucigel" may eventually screen completely the underlying epidermis surface of the root. The bacteria enclosed in the mucigel are in a favorable position for the utilization of soluble exudates passing from the roots towards the exterior. Further more, they are not in direct competition for the exudates with remainder of the complex rhizosphere population. The plant mucigel layer also contains sloughed-off root cap cells.[129,140] Generally, the sloughed-off cells retain their viability for 1 or 2 days[140] and are then subject to bacterial attack and degradation. They constitute a material with a C/N ratio probably close to that of average plant tissue, but quite different from the C/N ratio of the mucigel itself. From the point of view of their subsequent utilization by microorganisms, this difference may be important possibly favoring activities such as nitrogen fixation.

ii. Rhizoplane

In the basipetal parts of the roots, the mucigel layer, being more slowly produced by bacteria, soon disappears[138] and the epidermis itself becomes more susceptible to microbial attack. This part has been called the rhizoplane, or root surface, probably comprises bare epidermis, sloughed-off cells, enclosing mucigel, and in some cases also the endorhizosphere.

iii. Endorhizosphere

Bacteria and fungi enter the root tissues through holes they bore in the outer walls of the epidermis or through occasional wounds.[134,138] Such a microbial invasion may affect the entire cortex of a living root. This cortex with microbes is known as endorhizosphere. This population of the microbes is chiefly dependent upon the bacteria population in the mucigel. The microbial density is greater in the mucigel adjacent to the root cap cells as they decay very fast. These decaying cells provide the microflora with a substrate equivalent to the leaf litter so they have been called "root litter". The production of this root litter is related to the turnover rate of root. The roots of the tropical grass *Digitaria decumbens* cv. *transvaala* contain active N_2-fixing region of the roots with large colonies of N_2-fixing *Spirillum lipoferum*[141] which can be consistently isolated from surface sterilized roots of maize.[142]

iv. Rhizosphere Soil

Rhizosphere soil can be defined as that part of the soil in which the microbial activity is modified by the presence of the root and in which the microbial environment is mostly of soil origin. The flow of volatile and more or less soluble materials released by the plant crosses all of the compartments bed and reaches the neighboring rhizosphere soil. This is mainly responsible for sustaining microbial population. Further the microbial activities in this region can also have indirect effects such as change in pH and redox potential of the rhizosphere soil. As a result the microbial population keeps on varying. However, the rhizosphere soil is also rich in residues such as humic compounds, cell debris, and mucilage material which are relatively resistant to microbial attack.[132,143]

d. Nitrogen Fixation by Rhizosphere

Nitrogen-fixing bacteria are very frequent in the rhizosphere. Like other microorganisms, the microbial population in the rhizosphere must fulfill the two major requirements in order to fix the nitrogen. First, the *nif* genes must be present in a repressed state permitting synthesis of the nitrogenase system. Second, this system must also be supplied with electrons and with ATP; and since the enzyme nitrogenase is highly sensitive to oxygen, it must be protected from exposure to this gas. Physiological implications of these requirements were discussed by Postgate[99] and some of the general ecological implications for the free-living bacteria were discussed by Knowles.[128,144] Dobereiner[112] provided a comprehensive discussion of nitrogen fixation in the rhizosphere.

Many physical and biological factors influence the fixation of nitrogen in the rhizosphere and there seems to be no correlation between population size and nitrogen fixation. However, populations of 10^4 to 10^5 *Azotobacter* per gram soil are claimed to be responsible for significant nitrogen fixation in the rhizosphere of *Paspalum*[107] and in glucose-amended soil model system.[108] Further, at least 10^6 *Clostridium* per gram of soil are reported necessary for the fixation of substantial amounts of nitrogen in glucose-amended soil[108] and maize rhizosphere.[145] Another factor which markedly influences the nitrogen fixation in the rhizosphere is NH_4^+-N. The pool of NH_4^+-N must be sufficiently low to permit derepression of nitrogenase synthesis. Not much experimental data is available to support this because of the difficulty in the actual measurement of NH_4^+-N concentration in the rhizosphere. In a study of an anaerobic soil system, Knowles and Denike[146] found that threshold concentration for repression/derepression of nitrogenase activity was 4 μg NH_4^+-N per gram soil in the presence of 0.5% glucose whereas it was 35 μg NH_4^+-N per gram soil in the 1.0% glucose. There is also evidence from a marine carbonate sand that lower the available carbohydrate concentration, lower the NH_4^+-N concentration sufficient for the derepression of nitrogenase.[147] The plant itself acts as an ammonium sink, which takes up NH_4^+ directly by root and also indirectly through the stimulatory effect of high C/N exudates on the microbial proliferation and consequent immobilization of ammonium. The steep NH_4^+-N concentration gradients occur and this perhaps explains the observations of nitrogen fixation in the rice rhizosphere at a time when the average NH_4^+-N concentration in bulk rhizosphere soil samples was even 40 μg N/g.[139]

As mentioned earlier the rhizosphere microflora is supplied with carbon, energy, and reducing power in the forms of various types of gaseous, soluble and insoluble compounds released by the plant root. Most of these compounds are photosynthetic products like carbohydrates. These carbohydrates stimulate the nitrogenase activity and their absence reduces the nitrogenase activity. This was confirmed by reducing the rate of photosynthesis in seedlings[148] and by the occurrence of diurnal rhythms in nitrogenase activity.[139] The carbohydrates released may be utilized by nitrogen-fixing bacteria without competitions from the soil microflora. However, the efficiency of utilization in terms of nitrogen fixed per gram carbohydrate consumed in various soil, sediment, and pure culture system was reported to be in an inverse relation with the concentration of carbohydrates available.[149,150]

Indirect effects of the flow of exudates into the rhizosphere environment results in the depletion of NH_4^+ pool and increase in the consumption of oxygen by the microflora, thus creating an oxygen gradient. The existence of a gradient of oxygen concentration around plant roots is well documented in *Sinapis*.[151] *Sinapis* grows in somewhat aerobic soil and the concentration of oxygen drops markedly towards the root surface.[151] However, many plants have well developed lacunae which facilitate gas diffusion within the plant itself. Thus, rice[152-154] and many emergent plants such as *Spartina*[155] supply oxygen to the root so that root channels become oxidized and the oxygen gradient is therefore away from the root. It is likely, therefore, that each of the nitrogen-fixing bacteria isolated from the rhizosphere could find a zone in which the oxygen concentration is optimum for its development. Aerobes having an optimum O_2 (about 2 to 4%) such as *Azotobacter paspali*[112] *Azotobacter chroococcum*,[156] and *Spirillum lipoferum*[149] also frequently have greater efficiencies of nitrogen fixation (up to 30 to 50 mg N_2/g carbohydrates consumed) under such reduced concentrations of oxygen.[149]

Fixation of nitrogen by anaerobes and facultative anaerobes in the rhizosphere is restricted to anaerobic microsites which can be established readily in soil aggregates and on the surface of agar cultures.[157,158] *Klebsiella pneumoniae* although shows maximum activity under anaerobic conditions, can exhibit nitrogenase activity when grown in the presence of small concentrations of dissolved oxygen.[159] It should perhaps be stated here that in a study of a mixed-population soil model, O'Toole and Knowles[160] observed that the efficiency of nitrogen fixation not only increased as the available carbohydrate decreased, but was much higher in anaerobic than in aerobic soil. This may imply greater efficiencies of nitrogen fixation in the rhizosphere of plants growing on anaerobic substrates, but further studies are needed on this aspect.

Air humidity is another factor which can indirectly influence nitrogen fixing ability of microorganisms in the rhizosphere. This is presumably by inducing stomatal closure and thereby reducing photosynthesis of the plants which in turn will reduce the availability of carbohydrates. Low air humidity decreases nitrogen fixation activity in the rhizosphere of rice.[139] A related effect could be the reduction in diffusion of O_2 and N_2 through the plant tissues. It is thought that such an effect is probably responsible for the mid-day drop in nitrogenase activity sometimes observed *in situ* measurements.[139] The nature of soil in which a plant grows and water content of the soil[145] are examples of other important factors which influence directly or indirectly, nitrogen fixing activity in the root region.

A considerable extent of nitrogen fixation in the rhizosphere region occurs.[127,128,161] Reports on activities up to several thousand n-moles per gram dry root per hour exist and extrapolation of data on an area basis shows activities up to 300 g N_2 fixed per hectare per day. A report suggests that certain maize genotypes in Brazil can fix as much as 2 Kg N_2 per hectare per day.[142] It is interesting that rice, maize, and several tropical grasses are among the most active of the "terrestrial" systems. Significant quantities of nitrogen are also fixed in the rhizosphere of the freshwater emergent plants *Glyceria* and *Typha*,[162] the marine mangrove,[163] and the marine sea grasses *Zostera* and *Thalassia*.[147]

B. Symbiotic Nitrogen Fixation

The diverse group of bacteria that infect and produce nitrogen fixing nodules on members of leguminosae are placed in the genus *Rhizobium*. This genus represents organisms nodulating an enormous range of legumes occurring in diverse geographical and climatical areas. It may be expected that they are a heterologous group of organisms with differences significant enough to form more than one genus. Some rhizobia grow slowly while others are fast growers on certain laboratory media. The attempt to group all legume-nodulating organisms into one genus has caused confusion and much dissatisfaction in the classification of *Rhizobium*. The placement of all in one genus occurred during the first part of this century when only a small section of the family leguminosae had been examined. Recently, rhizobia were found to nodulate a plant, *Parasponia* (family ulmaceae), outside the legume family[164] which negates the original definition of *Rhizobium*. Present discussion is limited to genus *Rhizobium* and legumes in general.

1. Ecology of Rhizobia in Soil

The species of the genus *Rhizobium* are recognized on the basis of host preferences. Following are the main species which have been recognized.

Species	**Preferred host**
Rhizobium leguminosarum	*Pisum, Vicia, Lathyrus*
R. trifolii	*Trifolium*
R. phaseoli	*Phaseolus vulgaris*
	Phaseolus angustifolia
R. meliloti	*Medicago, Meliatus, Trigonella*
R. lupini	*Lupinus, Ornithopus*
R. japonicum	*Glycine max.*

In addition many strains of rhizobia which associate with cowpea and lotus are now recognized as cowpea rhizobia and lotus rhizobia.

Legume root nodule bacteria comprising the genus *Rhizobium* initiate with the help of an appropriate legume partner, a symbiosis of immense global importance to plant succession, soil fertility, agriculture, and general biological productivity. The rhizobia survive as free-living bacteria in the most complex natural soil environments. They are also capable of growth in the absence of host legume. A remarkable phenomenon occurs when the root of an appropriate legume is interposed into the environment of free-living rhizobia. It is a phenomenon of mutual recognition between the two compatible partners. The legume plant root, recognizes just the right kind of *Rhizobium* among all other bacteria, including other rhizobia in the vicinity and the *Rhizobium* in turn recognizes just the right kind of legume root among all the other roots which may occur in their environment.[165] The molecular bases for relative interaction between the partners are not known, nor is it known at what stage or stages in the interaction, specificity is expressed. Evidence that the interaction is underway is generally said to be found first in the elongation and deformation of legume root hairs. Long before that, many of the events that comprise the ecology of the rhizobia in soil and especially in the rhizosphere must be fulfilled in order to achieve successful interaction of symbiont and host.

There are compelling reasons to focus more of the attention on the ecology of *Rhizobium*-legume symbiosis in relation to the effects of pesticides. A major part of the pesticides, after all is applied to the soil which harbor the rhizobia-legume association. The effects of pesticides on such interactions can only be clearly understood when the ecology of rhizobia-legume associations is better understood. Questions about the survival, population densities, growth responses, interactions with other microorganisms, nutritional substrates, host specificity, and competitions between strains for host sites must be understood in the natural environment.

a. Rhizobia and Soil Physical-Chemical Features

The symbiosis of *Rhizobium* with legume is generally studied in soil under laboratory conditions. A few studies have been carried out under field conditions and there seems to be little information on the influence of environmental factors in natural soil environment. As mentioned earlier the understanding of such factors is of paramount importance in order to evaluate the actual toxicity of pesticides in terms of what may happen in natural environment.

i. Temperature

Rhizobia are mesophilic bacteria that function in subarctic, temperate, and tropical regions. Various kinds of rhizobia differ markedly in their tolerance to elevated temperature (around

40°C).[166] Rhizobia isolated from alfalfa showed considerable resistance to high temperatures, while those isolated from tropical hosts were quite variable in responses above their optimal growth temperature. The property of tolerance to high temperature of the inoculant strains is particularly significant in the introduction of forage legumes in semiarid subtropical region and during soybean cultivation in irrigated arid region.

Temperature response to the growth of rhizobia is also influenced by the other soil features. Marshall[167] observed that *Rhizobium meliloti* and *Rhizobium trifolii* were sensitive to high temperature in sandy soils, but not to the same temperature in heavier textured soil. This may partly be due to the surface properties of rhizobia which may lead to the formation of protective clay envelop resistance to high temperature and desiccation.[167] Gibson[168] studied the effect of root temperature on infection and nodulation. The possible effects due to soil colloids were excluded as the test plants and rhizobia were cultured on agar slants. The maximum temperature at which *Rhizobium trifolii* formed nodules on subterranean clover was 33°C and the minimum was about 7°C with most rapid initial nodulation taking place around 30°C. Initiation of nodulation slowed markedly below 22°C, and root hair infections were severely retarded above 33°C. Gibson[168] reported that low temperature adversely affected the infection and leghaemoglobin of early nodules in *Rhizobium trifolii,* but there were indications of low temperature tolerance and low temperature adaptation in other symbiotic system. *Rhizobium japonicum* is comparatively resistant to high temperatures.[167] Weber and Miller[169] also reported considerable variations among different strains in their ability to tolerate temperature. The strain of rhizobia found by Weber and Miller[169] were able to form nodules on soybeans at 30°C and formed very few nodules at 10 or 15°C. On the contrary some strains formed a few nodules at 30°C. Adaptation to low temperature by *Rhizobium trifolii* was clearly evident in isolates taken from subarctic regions of northern Scandinavia which grew faster and nodulated clover earlier when tested at 10°C than did isolates from southern Scandinavia.[170] There was little difference between the two groups when tested at 20°C.

ii. Moisture

The importance of water as an environmental factor in the ecology of soil rhizobia has been considered only from the point of view of survival. As the case for all soil bacteria, rhizobia must be surrounded by a water film in which the solutes are not so concentrated as to pose osmotic problems for the cell. Thus, too little water rather than too much is the threat to their survival and function. Extreme drying is accompanied by increased osmotic pressure in the soil solution, and if the desiccation is due to elevated temperature, this factor interacts as well. However, studies made only in relation to moisture contents without considering the interactions of other factors have no significance because of the interplay of soil factors and biotic effects during desiccation. The isolates of *Rhizobium trifolii* obtained from wet soils were poorly effective on test legumes as compared to those from dry sites.[171] Such an influence of soil moisture on the effectiveness pattern was probably a function of host plant rather than of rhizobia.

Moisture contents play a significant role on the survival of rhizobia and their subsequent infection is also evident from laboratory studies. Foulds[172] for example examined survival of rhizobia during air drying of 3 soils by plant infection techniques. *Rhizobium trifolii* proved to be more tolerant to draught than *Rhizobium meliloti* and *Rhizobium* of the lotus groups. An aspect of soil moisture that deserves additional attention is its impact on the movement of rhizobia. Soil moisture relative to soil pores would strongly affect the movement of the rhizobia. Hamdi[173] did extraordinary work on relationships between soil water tension and movement of *Rhizobium trifolii* from a point of inoculation into sterilized soil. With suction applied to soil of varying textures, it was demonstrated that the zones of movement were sharply decreased by increased water tensions, and movement was found to cease when water filled pores become discontinuous.

iii. Soil pH

There is general agreement that soil pH has a marked effect on the survival of rhizobia. As in the case of all other parameters, a great deal of variability in pH tolerance has been found among rhizobia. Soil pH in the acid range significantly affect the growth and survival of the bacteria rather than the symbiosis.[166] In reviewing some of the research on survival of rhizobia with respect to soil acidity, Vincent[166] concluded that *Rhizobium meliloti* is acid sensitive, and *Rhizobium trifolii* is less so, and *Rhizobium japonicum* is acid tolerant. Different strains within the same *Rhizobium* sp. also vary considerably in response to soil pH. Holding and King[174] found some strains of *Rhizobium trifolii* ineffective in nodulating the host but this ability was restored by the addition of lime which increases the pH.

Use of the quick serological testing procedure of Means et al.[175] made it possible to collect serotype of soybean nodules from a large number of sites. The same test was used by Damirgi et al.[176] in Iowa to identify the bacteriods found in field grown plants and examined the occurrence of serogroups with respect to soil properties. Serogroup 123, 135, 31, and 3 included nearly all *Rhizobium japonicum* bacteriods. Serogroup 123 dominated in soils with a pH of 5.5 to 7.5, while serogroups 135 was dominant in the highly alkaline soils. More detailed work by Ham et al.[177] confirmed the overall importance of soil pH in the dominance of these serogroups, but indicated that other soil factors should also be considered. In the Iowa studies mentioned above, numerous serogroups were represented in each of the soils, and although serogroups 123 occupied most of the nodules in many soils, the proportion of subdominants varied considerably from soil to soil.[176] The nodule serotype of the indigenous rhizobia is difficult to mention with certainty as the related factors which also effect the populations of rhizobia were not considered.

The practice of introducing appropriate rhizobia into the soil by way of inoculated seeds of host legumes has been common since the beginning of the century. With the development of a nodulated crop, large numbers of the newly introduced rhizobia are released into the soil to join the complex indigenous microbiota already there and in equilibrium with the soil environment and each other. It is particularly significant that most of the rhizobia along with bacteria are added to the rigorously competitive, energy limited life. However, some of them may not tolerate transition from luxurious competition free existence. Some of the most practically important questions in ecology of rhizobia center on this adaptation of rhizobia. The example with respect to *Rhizobium japonicum* is illustrative. Soybeans and *Rhizobium japonicum* were first introduced into soils of midwestern USA perhaps 50 years ago; both are now widespread in the west soils of the region. When Ham et al.[177] examined some of these soils, consistently they found that many different strains of *Rhizobium japonicum* (those comprising serogroups 123) appeared most frequently in the field nodulated plants. If the 123 serogroup predominated by virtue of its superior adaptation to a saprophytic soil existence, which is a distinct possibility, how has this strain become so successful? Can the elements of success for 123, if once discovered be applied to the introduction of other strains, other than *Rhizobium* sp. — for that matter, other bacteria. Though none of the many possible ingredients of eminent success as a serotype have been resolved for even one strain, an obvious essential is an adequate carbon and energy substrate in the soil. What the substrates are and what level of growth the rhizobia attain in competition with all other microbiota is an important mystery.

b. *Rhizobia and Soil Biological Features*

A very wide range of biological interactions both in terms of mechanisms and biotic participants have been postulated in the literature as of potential importance to rhizobia in the soil or in the rhizosphere. A clear knowledge of these interactions can certainly help a lot while evaluating the effects of pesticides on rhizobia. Because of the limitations imposed by techniques available to deal with the vast complexities of the soil, little work relates directly to natural environment and few specific interactions have been resolved. Interest in soil rhizobia as participants in biological interactions has stemmed largely from practical problems such as nodulation failures in some soils and widely reported difficulties in the introduction of favored strains into an established population of indigenous rhizobia. In view of this it is not surprising that the emphasis has only been on inhibitory and antagonistic effects.

i. Microbial Interactions

Microbial antagonism by fungi and bacteria has been demonstrated repeatedly against the rhizobia in laboratory culture or sterile soil. Experiments of this type have established that a great many pure culture isolates taken from soil or rhizosphere and grown in rich culture media can produce antagonistic (frequently antibiotic) effects against rhizobia. Extrapolation of these to the complex, competitive, low-nutrient soil situation is very difficult as few investigators have studied the potential importance of this aspect in natural environments.

Holland and Parker[178] tested 286 bacteria and fungi for antagonistic effects on *Rhizobium trifolii*. *Phoma* and *Aspergillus* were inhibitory while *Penicillium* and two phycomycetes were bactericidal in laboratory assays. All but two of the penicillia were isolated from areas where nodulation failures were common due to antagonistic effects. The isolates were grown aseptically in tubes of sterilized sand, together with inoculated subterranean clover. Of the 31 cultures that inhibited nodulation in sand cultures, 29 were strains of *Penicillium*. Even in these artificial conditions of a noncompetitive environment, the penicillia interfered with nodule formation only when sucrose and straw were present at relatively high concentrations. Holland and Parker nevertheless considered it likely that *Penicillium* was responsible for the failure in nodulation in some areas. Interactions that may occur in soil between rhizobia and common soil fungi such as *Penicillium* are further complicated by the lack of any standard experimental guidelines. Sporulating fungi appear readily on dilution plates and are easily isolated, but the dilution plate colony may represent merely a spore that was quiescent in the soil and hence incapable of significant antagonistic influence. Holland and Parker[178] further found that the aqueous extracts from the soils with nodulating plants were inhibitory to *Rhizobium trifolii* on seeded agar plate indicating the presence of "antibiotics", but unfortunately, the activity of the extracts was lost before, the antibiotic nature of the antagonism was demonstrated.

Parker and Grove[179] found a predatory bacterium *Bdellovibrio bacteriovorus* in six out of the ten soils which was able to lyse *Rhizobium meliloti* and *Rhizobium trifolii*, while *Rhizobium lupini* was not lysed by any of the strains of *Bdellovibrio*. In addition, two species usually considered as close relatives of the rhizobia, *Agrobacterium tumefaciens* and *Agrobacterium radiobacter*, were also lysed by *Bdellovibrio*.

ii. Bacteriophages and Bacteriocins

Bacteriophages specific for rhizobia apparently are common in soils and nodules. The possibility that bacteriophage could have deleterious effects on rhizobia was first mentioned more than 40 years ago. These bacteriophages are generally, lytic to host strains of *Rhizobia japonicum*.[180] Vincent[166] concluded that despite a relatively voluminous literature, the practical significance of bacteriophage against the survival and functioning of rhizobia in the soil has yet to be unequivocally demonstrated. In recent years transfer of genetic information in the form of plasmids to rhizobia have been discovered by means of bacteriophages. In the light of this it becomes important to trace the bacteriophages in the soil and rhizosphere to evaluate their practical potential. Another role of plasmids in rhizobia is the production of bacteriocins. Bacteriocins are a diverse group of substances, composed largely of proteins, produced by many bacteria which kill certain other bacteria on a highly selective basis. Bacteriocins are strain specific and are thus distinguished from antibiotics which have a wider spectrum of activity. The ability to produce bacteriocins is often inherited via plasmids. Roslycky[181] was the first to observe bacteriocins in rhizobia, and noted their occurrence in several cross-inoculation groups. Bacteriocins of *Rhizobium lupini* were characterized by Lotz and Mayer[182] as defective bacteriophage particles since they resembled the tails of certain *Rhizobium* and *Escherichia coli* phage, but had no nucleic acid. Schwinghamer et al.[183] pointed out the ecological significance of bacteriocins. They stated that bactericidal function of the bacteriocin could be significant in providing the bacteriocinogenic strain with some competitive advantage over other rhizobia.[184]

c. Persistence of Rhizobia in Soils

The interaction of pesticides with rhizobia becomes more complicated when studies are to be carried out in the field. It is further complicated when persistence strains of rhizobia with a different nature are added to the soil. If the soil has an indigenous population of rhizobia or if it is necessary to use a mixed inoculum, it is essential to differentiate the strain to be enumerated from other strains that may appear in the nodule. It has not been possible to investigate the persistence of a given strain relative to one or more competing strains in the same environment because of the limitations imposed by the selectivity of the plant infection technique of enumeration. This difficulty has been recently solved to some extent with the advent of immunofluorescence techniques.[185-189]

Dudman and Brockwell[190] used immuno-diffusion techniques to examine the persistence in the field of two strains of *Rhizobium trifolii* supplied as commercial peat inoculant to clover seed. They looked for the strains at intervals of 3 to 42 months. One strain dominated the other but both diminished greatly with time. At one site, one inoculum strain was recovered after 30 months, but at another it had disappeared by 18 months. Such a behavior of the different strains of rhizobia has to be studied and considered in detail when making any conclusion on the toxicity of pesticides on microorganisms. The presence of natural population had little to do with the ability of the inoculant strain to establish in soil. The survival of field grown rhizobia in certain problem soil of Western Australia was studied by Chatel and Parker.[191] Their findings suggested that colonization during the growing periods was critical in order to build up sufficient numbers to withstand long hot dry summer. In further experiments designed to examine colonization during crop development, Chatel et al.[192] observed marked difference in survival patterns between *Rhizobium lupini* and *Rhizobium trifolii* during the second year.

The site of colonization in the soil is of central importance to the persistence of rhizobia, and so long as host legume cultivation is continuous, the colonization site is provided in and around the host roots. The interesting question of dependence of rhizobia on particular microenvironments in the soil mass was examined by Chatel and Greenwood[193] who looked at soil cores in senescenced legume pastures. They divided soil cores into soil, host plant roots, nodules, and extraneous material fractions for estimates of populations of *Rhizobium trifolii* and *Rhizobium lupini*. The former was found to be predominately associated with the nodule fraction of soil cores, while the latter was distributed more uniformly in cores from beneath senescenced soil. Unfortunately, the subsequent persistence of these rhizobia relative to their respective fractions was not included in the study. Because commercial inoculant strains did not colonize soil and persist there, Chatel et al.[192] included fresh field isolates with inoculant strains to evaluate persistence. The fresh isolates were originally selected for superiority in efficiency as well as persistence and in particular they showed much improved persistence over the older inoculant strains. In all the soils studied, the live cells persisted for 2 to 4 months although their number reduced drastically with the passage of time. This study demonstrated the applicability of the persistence of a particular strain of *Rhizobium* in soil amid other rhizobia. The above study was only semiquantitative as it was based on comparative counts of buried slides and the methods for quantitative fluorescent antibody examination of soil were still not fully developed. The first report on the fluorescent antibody method for enumeration of soil bacteria was illustrated by data on *Rhizobium japonicum* 110.[194,195] The study also included short-term persistence of the *Rhizobium* strain 110 in two soils. This time strain 110 was almost double in number but the generation time was extremely slow: 241 hr in the Clarion soils and 361 hr in the Ulen soil as compared to 14 hr in autoclaved, sterile Clarion soil. Thus, soil colonization and persistence abilities are very important aspects in the ecology of the rhizobia in soil. Without the understanding of these parameters any conclusion regarding the toxicity of the pesticides will not be of much significance.

i. Rhizosphere Effect

Rhizosphere, as mentioned earlier is the zone near the root surface which shows high intensity of microorganisms. The basis for this high microbial population are root exudates and cellular debris contributed by the plant root. The high population of microorganisms in the rhizosphere as compared to the remaining neighboring soil is known as the rhizosphere effect. Depending on the nature of the plant there is always diversity in the amount of root exudate available. The legume rhizosphere nearly always varies depending on the amount and diversity of substrate materials contributed by the root exudate. However, *Rhizobium* population also varies greatly according to the manner in which data are expressed.[196] Legumes exhibit an impressive rhizosphere effect when the microbial population is expressed on the basis of weight of rhizosphere soil, but are greatly outstripped by grasses when the rhizosphere count is expressed on the basis of root surface. The rhizosphere effect is nonetheless real in the legume, and the rhizosphere zone is overwhelmingly important to the legume rhizobia symbiosis. The soil factors which influence the population of microbes are also present in the rhizosphere and these factors will certainly exert some influence. However, it is unfortunate that the ecology of rhizobia in relation to legume rhizosphere is poorly understood. The following is a brief discussion on the nature of *Rhizobium* population in such environments.

There is a qualitative and quantitative difference between the microbial population of soil and rhizosphere. Any bacterium present in soil cannot enter into legume rhizosphere.[197] The exact nature of interactions when the legume seed germinates and rhizosphere development begins is unknown, but the rhizobia find suitable substrates and compete effectively for them. The establishment of rhizobia in the vicinity of the germinating seed may be delayed by inhibitory diffusates from the seed coat.[166]

The formation of legume rhizosphere provides environments suitable for the existence and adaptation of rhizobia. This is evident from experiments in which numbers of rhizobia capable of nodulating a test legume was compared in rhizosphere soil with nearby nonrhizosphere soil. Tuzimura and Watanabe[198] found 10^3 to 10^4 per gram soil rhizobia that nodulate *Astragalus* in the legume rhizosphere but adjacent nonrhizosphere soil in the same pot had rhizobia less than 10^1 per gram soil. Rovira[199] counted *Rhizobium trifolii* from red clover rhizosphere and nonrhizosphere soil using soil dilution technique. Rhizobia in soil unaffected by plant roots numbered about 2×10^5 per gram, whereas numbers in soil adhering to the clover roots increased to 200-fold. Further work by Tuzimura and Watanabe[200] demonstrated that *Rhizobium trifolii* increased 10- to 100-fold in the rhizosphere of legumes. Substantial rhizosphere effects for *Rhizobium meliloti* and *Rhizobium leguminosarum* were recorded by Nutman[201] and from legumes by Krasilnikov.[202] The increase in number of rhizobia as a result of extensive multiplication in legume rhizosphere is essential to favor a particular kind of *Rhizobium* that can enter into symbiosis.[201,203,204] It is of particular interest to mention here that only the *Rhizobium* which has to enter into symbiosis multiply rapidly and increase in number. However there is no data to explain this specific event. Rhizobia counts in the rhizosphere in relation to total bacteria counts indicate that the rhizosphere effect for rhizobia is due to the "selective stimulation" of that particular *Rhizobium*.[198-200] What is responsible for the stimulation is not known.

The concept of specific stimulation of a particular *Rhizobium* by the rhizosphere of its host legume stems from rhizosphere effect. Nutman[201] citing the studies of Krasil'nikov[202] that indicated a greater density of *Rhizobium trifolii* in the rhizosphere of clover than in other legumes concluded that there was large and specific stimulation and that "a given legume tends to promote the multiplication of bacteria able to nodulate it more than others". A few attempts seem to have been made to compare populations of rhizobia of different inoculation groups in the same legume rhizosphere or the populations of different strains of the same *Rhizobium* in the rhizosphere of compatible host. This is necessary to establish the existence of a specific stimulatory effect because legumes other than the host legume make rhizospheres suitable for the growth of a given *Rhizobium*. This was seen in the experiments of Tuzimura and Watanabe[200] wherein *Rhizobium trifolii* was only slightly enhanced in the rhizosphere but exhibited a large rhizosphere effect with alfalfa and soybean neither of which can be nodulated by *Rhizobium trifolii*. There is also consistent evidence that rhizobia display definite rhizosphere effects with a wide range of nonlegume plants. Rovira[199] demonstrated enhanced development of *Rhizobium trifolii* in the rhizosphere of the grass *Paspalum*. Tuzimura and Watanabe[200] reported marked rhizosphere effects of *Rhizobium trifolii* with rap and tomato, and smaller effects for rice, wheat, and sudan grass. Both *Rhizobium trifolii* and *Rhizobium meliloti* established better in the rhizospheres of a native Australian grass.[205] Rhizobia do so well in the rhizosphere of certain nonlegumes which could be a factor favoring survival of rhizobia in the absence of host plants.[196] *Rhizobium japonicum* can be introduced into the soil by inoculating seeds of cereals and nonhost legumes.[206] But this approach is only successful in the greenhouse and much less so in the field conditions. Thus, the question of specific stimulation of the rhizobia in the rhizosphere remains unresolved as the population of rhizobia is also stimulated even in the absence of host plant by nonhost.

Inoculation of soybeans with selected strains of *Rhizobium japonicum* has been notably unsuccessful in influencing nodulation or enhancing nitrogen fixation in regions where soybeans have been cultivated previously. The inoculant fails to nodulate the soybean to any significant extent, whereas virtually all of the nodules are occupied by strains of *Rhizobium japonicum* indigenous to the stock.[177,207-209] Throughout much of the major soybean production area of the U.S., indigenous soil strains of *Rhizobium japonicum* serogroup 123 dominate nodulation of soybeans regardless of the inoculation practice or of the cultivar grown.[176,177,188,210] Other serogroups of *Rhizobium japonicum* also occur as indigenous soil bacteria, presumably introduced during previous soybean culture, but these indigenous strains are no more successful than the strains added as the inoculant. The success of particular indigenous strain may be contigent on competitive interactions with other strains in the host rhizosphere before nodulation.[185]

The advent of immunofluorescene techniques has made it possible to examine and to study the autecology of individual serogroups of *Rhizobium japonicum* directly in soil and in natural rhizosphere.[185,187-189,194,211] Schmidt and co-workers have recently made use of serogroup specific fluorescent antibodies to follow the population dynamics of *Rhizobium japonicum* serogroup 123 and two competing serogroups of *Rhizobium japonicum* within the same rhizospheres of field-grown plants during the course of two growing seasons.[186] Populations ranged from low densities of several thousand per gram in the early spring to high densities approaching million per gram in the early fall. Serogroups 123 was estimated to compose from 10 to 50% of the *Rhizobium japonicum* population thus dominating the field grown soybeans as was anticipated from several earlier investigations.[188,189,212] This dominance was consistent each year for the cultivar examined and it was not changed materially either by inoculation with *Rhizobium japonicum* of another serogroup or by plant

Table 5
POPULATION DENSITIES OF THREE INDIGENOUS SEROGROUPS OF *RHIZOBIUM JAPONICUM* IN HOST AND NONHOST RHIZOSPHERE OF FIELD-GROWN PLANTS DURING THE 1980 SEASON

Density (No. × 10⁶/g) soybean cultivar

Week after seeding	Chippewa			70—017			70—017[a]			Nonhost oats		
	110	123	138	110	123	138	110	123	138	110	123	138
1	0.4	0.5	0.9	0.2	0.4	ND	0.1	0.1	ND	0.1	0.1	0.1
2	0.9	0.8	1.0	0.6	1.0	0.9	0.8	1.1	0.8	0.4	0.2	0.3
3	0.5	1.0	1.1	0.2	0.7	0.3	0.6	0.5	0.5	0.1	0.1	0.1
4	0.7	1.1	1.5	0.6	0.7	0.6	ND	ND	ND	ND	ND	ND
5	2.0	2.2	1.6	1.7	1.9	2.5	2.5	3.0	4.6	0.1	0.3	0.3
9	44	1600	270	130	5500	220	78	810	4.3	1.3	4.0	4.7
15	6.1	1600	15	3.3	1800	5.2	20	6900	260	0.3	0.3	0.7

Note: ND = Not determined.

[a]   Seed inoculated with *Rhizobium japonicum* serotype 110 at planting.

From Moawad, H. A., Ellis, W. R., and Schmidt, E. L., *Appl. Environ. Microbiol.*, 47, 607-612, 1984. With permission.

maturity. As the nodule occupancy by serogroup 123 was expressed in the presence of serogroup 110 and 138, the success of serogroup 123 could not be attributed to the methodology employed. Further, all the serogroups of all indigenous population were present in the soil at about the same cell density, so ultimate success of the serogroup 123 could not be attributed to a numerical predominance at the time of planting. The success of serogroup 123 was again attributed to rhizosphere effect of selective stimulation. This is supported from the facts that early development of seedlings was usually accompanied by enhanced growth of *Rhizobium japonicum* in the rhizosphere as compared with that of fallow soil.

However, the rhizosphere effects were slight, no greater for serogroup 123 than for its competitors.[186] The pattern of host rhizosphere colonization by three populations of competing *Rhizobium japonicum* was similar for each of the two growing seasons (Tables 5 and 6). Rhizosphere effects were slight for newly emerged soybeans, increased gradually with increased root development before flowering, and then declined during plant maturity (Table 7). Rhizosphere populations especially of serogroup 123, were still higher during pod fill and seed maturation. Moreover, the three groups combined made up only a small minority of the rhizosphere bacteria. These data are consistent with that of rhizobia in early rhizosphere of host plants observed for soybeans[188] and for *Phaseolus vulgaris*.[187] Thus, the extensive multiplication of microsymbiont in the host rhizosphere is not a prelude to the initiation of nodulation as has been postulated previously.[123] The factors responsible for high populations of indigenous *Rhizobium japonicum* in the rhizospheres of host plants during flowering as reported by Moawad et al.[186] are not clear. It is possible that all three serogroups and especially 123 were responsive to the rhizosphere conditions imposed at this active stage of plant development. Such pronounced host rhizosphere effects may be a mechanism whereby diverse *Rhizobium japonicum* strains maintain a population base in soil despite their inability to compete successfully in nodulation. The ability to colonize nonhost rhizosphere, as in oats may also serve to maintain indigenous *Rhizobium japonicum*, though possibly at lower population densities in the absence of host crop.[186]

Specific stimulation hypothesis was supported by the experimental data of Egeraat.[213] He

Table 6
POPULATION DENSITIES OF THREE INDIGENOUS SEROGROUPS OF *RHIZOBIUM JAPONICUM* IN THE RHIZOSPHERE OF FIELD-GROWN SOYBEAN CULTIVARS DURING THE 1981 SEASON

	Density (No. $\times 10^6$/g) of cultivar and serogroup											
	Chippewa			Chippewa[a]			70—017			70—017[a]		
Days after seeding	110	123	138	110	123	138	110	123	138	110	123	138
13	0.1	0.3	0.2	0.7	0.6	0.6	0.2	0.9	0.2	4.7	0.2	0.1
18	0.7	1.1	0.4	3.9	2.3	0.8	0.6	1.0	0.5	2.4	0.7	0.2
22	0.4	0.9	0.3	1.2	1.1	1.6	0.8	1.5	0.3	5.2	8.3	1.1
35	0.5	1.3	0.6	0.8	1.7	0.9	0.5	1.0	0.7	0.5	2.3	0.5
48	3.1	200	6.9	9.1	31.6	4.9	13.8	75.9	13.5	7.2	224	10.7
53	1.9	12.6	2.9	1.1	6.8	0.4	0.3	6.8	0.3	72.4	191	3.0
82	0.5	17.4	0.4	0.2	2.6	0.1	0.2	2.8	0.1	0.9	28.2	0.2

[a] Seed inoculated with *Rhizobium japonicum* serotype 110 at planting.

From Moawad, H. A., Ellis, W. R., and Schmidt E. L., *Appl. Environ. Microbiol.*, 47, 607—612, 1984. With permission.

Table 7
COMBINED POPULATIONS OF INDIGENOUS *RHIZOBIUM JAPONICUM* SEROGROUPS 110, 123, AND 138[a] AS A PERCENTAGE OF TOTAL BACTERIA[b] IN 1980 AND 1981 FIELD PLOTS

	% of total bacteria as indicated week after planting			
Cultivar rhizosphere or fallow	1980		1981	
	1	15	3	12
70 — 017[c]	0.003	20.0	0.6	0.02
70 — 017	ND	1.8	0.2	0.02
Chippewa[c]	ND	5.2	0.7	0.2
Chippewa	0.01	2.7	0.2	0.9
Fallow	0.001	0.001	0.05	0.007

Note: ND = not determined.

[a] Total of specific FA counts.
[b] Total bacteria as determined by acridine orange direct microscope count on membrane filters.
[c] Seed inoculated with *Rhizobium japonicum* serogroup 110 at planting.

From Moawad, H. A., Ellis, W. R., and Schmidt, E. L., *Appl. Environ. Microbiol.*, 47, 607-612, 1984. With permission.

studied the root zone of sterile pea plants in solution culture in attempts to identify root exudates that influence the pea symbiont *Rhizobium leguminosarum*. He conducted detailed analysis of ninhydrin-positive compounds excreted by pea plants under the above conditions to find that homoserine component was quantitatively most important. In testing growth responses to homoserine in pure culture it was found that *Rhizobium leguminosarum* grew very well with the amino acid as the only carbon and nitrogen source, whereas homoserine was toxic to the strains of *Rhizobium trifolii* and *Rhizobium phaseoli*. This indicates that there are appealing mechanisms whereby appropriate pea rhizobia are favored while at the same time rhizobia of other inoculation groups are repressed.

ii. Mobility of Rhizobia in the Rhizosphere

The contact mechanisms of a given *Rhizobium* with the appropriate part of an appropriate legume root system to commence a symbiosis is still not fully understood. In most of the studies only the possibility of enhanced growth in the rhizosphere and consequent multiple contact with growing root system have been considered. The mechanism of covering distance quite away from the legume roots to make contact with the legume roots has been considered by Hamdi.[173] However, Hamdi[173] considered only the effect of moisture tension on rhizobial migration in sterilized soil. Along with the needed studies of rhizobial movement in the rhizosphere from a point of inoculum as a function of growth rates, consideration should also be given to means alternative to growth. Chemotaxis is one such biological mechanism that could conceivably account for the movement of a specific bacterium from other rhizosphere to a specific site on the host root, and as well as for the selectivity of the host in choosing one strain among many.

Dart and Marcer[214] made an electron microscope study to find out the structural changes in the rhizobia and root at the time of contact. Barrel medic seedlings were grown in sterile solution culture and inoculated with *Rhizobium meliloti*. Under these highly artificial "rhizosphere" conditions the authors obtained electron micrographs some of which showed "swarmer" cells among the microfibrils of the root hairs. These cells were coccoid and multiflagellate, some of them as small as 0.1 μm in diameter. The possibility that such cells were the infective stage of *Rhizobium,* and capable of reaching nodulation sites by virtue of extreme mobility was suggested.

iii. *Rhizobium* Host Root Specific Recognition

Rhizobia, including those ultimately successful nodules, probably are always outnumbered by other legume rhizosphere bacteria given even the relatively high densities. Just how the recognition takes place between a host legume root and one specific kind of *Rhizobium* amid all the activity of the rhizosphere is a question of immense interest. An answer to this question for example, could be a key to the realization of the old but honorable dream of extending the nitrogen-fixing symbiosis to nonlegumes of agricultural interest. Extensive study of a root hair infection occurs at specific sites on the root system. Thus, a recognition mechanism of some sort must be called into play at the root hair infection site soon after infection thread formation begins.

Few hypotheses have been advanced to account for the specificity of the *Rhizobium*-legume association and only one has generated data at the biochemical and physiological level. In this it was suggested that the specificity is governed by the ability of the bacterium to induce the enzyme polygalacturonase in roots of compatible hosts.[215] This enzyme was suggested to cause softening of the cell walls of root hairs, preliminary to invasion by the *Rhizobium*. This hypothesis seems to not hold, nor does it account for the specificities expressed at the strain level to explain the manner in which particular strain induces the mobilization of a relatively nonspecific enzyme.[216] It is a matter of common observation that with a large number and perhaps numerous kinds of rhizobia in a legume rhizosphere the recognition of an appropriate *Rhizobium* partner would occur at the infection site and not after an infection thread had started. In that case the appropriate *Rhizobium* must carry some recognition factor as a rhizosphere bacterium that differentiates it from other rhizobia and from other bacteria to make it recognizable to the plant. Hubbell[217] looked for such a marker and obtained a crude extracellular polysaccharide preparation from a *Rhizobium* sp. that mimicked the root hair curling involved in infection in strawberry clover.

A new hypothesis that legume-rhizobia specificity is mediated by lectins produced by the legume has also been advanced.[218,219] Lectins, sometimes referred to as phytohaemagglutinins, are cell agglutinating, sugar-specific glycoproteins found widely in legume seeds. According to this hypothesis the lectins present on the legume root surface interacts specifically with a distinctive polysaccharide on the surface of the appropriate *Rhizobium* cell as a prelude to nodulation. Hamlin and Kent[219] using only a single strain of *Rhizobium phaseoli* found that lectin derived from the bean *Phaseolus vulgaris* bound to bacteria of that strain. Evidence for lectin binding was that the lectin-treated bacteria were capable of aggilutinating red blood cell — a characteristic property of lectins. Control cells of *Rhizobium phaseoli* had no hemagglutinating activity. Whereas these experiments did not address the question of specificity, those of Bohlool and Schmidt[218] sought to examine this point. They devised a technique whereby reactions between soybean lectin and *Rhizobium japonicum* could be studied by direct microscopy. It was found that soybean lectin labeled with fluorecein isothiocynate as a marker combined specifically with all but 3 of 25 strains of *Rhizobium japonicum*. The lectin did not bind to any of 22 other strains representative of rhizobia that do not nodulate soybeans.

Lectins have been known for many years by virtue of their ability to react with erythrocytes and cause their agglutination, but the role of lectins in nature has been a mystery. Hemagglutination activity, as well as certain other biological properties of the lectins, perhaps have nothing to do with their function in nature: still, two properties fit well with a hypothetical role for lectins as the determinant of specificity in the legume-rhizobia interactions. One property is binding specifically to saccharides on the surface of cell; this has provided a new tool for the investigation of the architecture of cell surface[220] and makes it entirely reasonable to conceive of legume lectins combining specifically with small saccharide-binding sites on the surface of appropriate *Rhizobium* cells. The second property, the mitogenic action of lectins, though less obvious is worth consideration nevertheless. If lectins are localized at actively growing sites of root extension, i.e., the nodulation sites, they may function there as mitogens to increase the rate of cell division. The same lectins if bound to the rhizobia during infection could conceivably be carried to the cortical cells acting as stimulus to the cell division that initiates nodule primordia.[221]

The hypothesis that legume lectins mediate the specificity of the *Rhizobium-legume* interaction is most attractive although the evidence in its support is still preliminary. It may be anticipated that considerable physiological and biochemical work needs to be carried out on the role of lectins in rhizobia-legume symbiosis. Ecological studies bearing on the lectin hypothesis will be difficult unless the complexities of the natural environment are avoided as has been done so frequently in ''rhizosphere'' studies. The availability of fluorescent antibodies for the detection of specific rhizobia in the natural soil environment provides the expectation that localization and perhaps even orientation of rhizobia at hypothetical lectin binding sites on roots may be visualized.[222]

iv. Competition

The same factors that enter into the capacity of a *Rhizobium* to establish in a soil in competition with the biota already present, come into play more intensively in the rhizosphere and are augmented there by additional factors. Away from the root in the nonrhizosphere soil, competition is likely to revolve around the ability of a *Rhizobium* to utilize a given

Table 8
IDENTIFICATION AND SEASONAL
VARIATIONS OF *RHIZOBIUM JAPONICUM*
SEROTYPES IN NODULES OF SOYBEANS
INOCULATED WITH WAUKEGAN SOIL

	% Serogroup							
	123		138		110		Others	
Sampling time	S	M	S	M	S	M	S	M
7/79	50	12	8	24	0	14	14	6
9/79	44	12	24	4	12	8	4	8
12/79	43	11	7	7	11	14	14	11
4/80	22	16	28	16	6	0	25	0

Note: S = single infection; M = mixed infection.

From Moawad, H. A., Ellis, W. R., and Schmidt E. L., *Appl. Environ. Microbiol.*, 47, 607—612, 1984. With permission.

substrate rivalry with other bacteria. In the rhizosphere, competition for substrate seems to go at least somewhat in favor of the rhizosphere, rhizobia must compete with other rhizobia not only for available substrate but for nodulation sites as well. This latter consideration has dominated studies of competition because of the practical importance of evaluating a good inoculant strain in terms of its ability to form nodules in competition with nodulating strains already established in the soil. In the competition for nodulation sites, the legume plant plays a selective role that is not understood but which needs to be given much greater recognition as a factor in competition studies.

A strain that is to be studied in a normal rhizosphere with other soil microorganisms (and usually other rhizobia), must be identifiable. Identification has usually been done by immunological techniques, but the work of Means et al.[175] is an interesting departure. They used a *Rhizobium japonicum* strain marked by its ability to induce a chlorosis in certain soybean cultivars and found that it was highly competitive in mixtures with other strains. Mixtures containing as little as 1% of the chlorosis-inducing strain resulted in that strain forming 85% of the nodules.

A good deal of the work on strain competition has centered on *Rhizobium japonicum*.[208] The extent to which an inoculant strain can compete with resident rhizobia undoubtedly is a function of soil conditions, the natural competitors, the legume, and soil colonizing ability of the introduced strain. Bohlool and Schmidt[211] plotted the percent nodules against inoculant densities. A sigmoidal curve was obtained between percent nodules due to an inoculant strain vs. the log number of the inoculant over a range of inoculant densities.[211] Such a competition curve was proposed as a possibly useful means to evaluate the interacting factors for a particular soil and legume combination.

The competitive ability of a strain of *Rhizobium* has been evaluated only in terms of competition for nodulation sites, since the nodule is a prerequisite to strain recognition. In order to evaluate the selective barrier effect of the host and to gain an understanding of rhizosphere factors that influence competition it is necessary to augment the nodulation site approach with a more direct study of the behavior of a strain in the rhizosphere. Moawad et al.[186] found that among the commonly available serogroups (123, 138, and 110) only 123 thrived in the competition and dominated nodulation of soybeans. Under the condition of growth pouch analysis, nodules from most soil inocula were occupied by serogroup 123 (Table 8). Serogroup 138 was the second most commonly occurring nodule occupant followed

Table 9
**RHIZOBIUM JAPONICUM SEROGROUP OCCUPANCY OF EARLY
(5 WEEKS) AND LATE (12 TO 15 WEEKS) NODULES ON SEVERAL SOYBEAN
CULTIVARS STUDIED IN THE 1980 AND 1981 FIELD PLOTS**

| | | % of serogroup present in nodules | | | | | | | |
| | | 123 | | 138 | | 110 | | Others | |
Cultivars	Inoculant	E	L	E	L	E	L	E	L
Chippewa	None	84(95)	72(93)	12(0)	12(18)	4(3)	10(10)	12(5)	14(5)
	110	70(90)	78(100)	14(5)	4(0)	14(10)	24(0)	18(10)	26(0)
	138	(75)	(90)	(0)	(10)	(45)	(30)	(15)	(0)
Hodgson	None	70(93)	70(88)	8(10)	8(3)	12(3)	22(10)	26(8)	16(13)
	110	64(90)	62(95)	4(0)	14(20)	20(25)	6(5)	24(5)	38(0)
	138	(95)	(90)	(20)	(0)	(5)	(25)	(0)	(5)
70-017	None	84(98)	78(100)	6(15)	6(5)	6(5)	10(13)	18(3)	38(0)
	110	82(90)	88(95)	6(0)	6(20)	14(25)	12(5)	18(5)	16(0)
	138	(95)	(90)	(20)	(0)	(5)	(25)	(0)	(5)

Note: Values listed report the serogroup present in both single and mixed infection nodules as determined by immunofluorescence. Mixed infection frequency in 1980: serogroup 123, 14.5%; serogroup 110, 6.3%; serogroup 138, 2.0%. Seeds inoculated with designated strain of *Rhizobium japonicum* at time of planting. E = early, 5-weeks nodules; L = late, 12- or 15-weeks nodules. Numbers in parentheses report percentages of nodules occupied by a given serotype in 1981; number not in parentheses are for 1980.

From Moawad, H. A., Ellis, W. R., and Schmidt, E. L., *Appl. Environ. Microbiol.*, 47, 607—612, 1984. With permission.

by serogroup 110. Table 9 shows the serogroup occupancy of early and late nodules on several soybean cultivars during two growing seasons.[186] In this case also serogroup 123 was dominant during both the years. However, dominance of serogroup 123 with respect to nodule occupancy was quite clear in field grown nodule but was not as clear cut in pouch grown nodules. Serogroup 123 was less competitive by pouch analysis, whereas others (serogroup 138 for example) appeared much more competitive in pouches than in soil. Reasons for the discrepancies are not known, but clearly competition in the growth pouch did not mimic the natural environment.

d. The Cross-Inoculation Group

Early studies on the legume-*Rhizobium* symbiosis were concerned with species such as those belonging to *Pisum, Vicia, Lens, Lathyrus, Trifolium,* and *Medicago.* It was soon shown that the isolate from one genus would not necessarily nodulate, in another and that the legumes studied formed groups based on the ability of their associated rhizobia to be mutually interchangeable. The rhizobia nodulating each group were generally named according to one of the hosts of that group (Table 10). These cross-inoculation groups were developed on the premise that each species of *Rhizobium* would only nodulate plants within the particular "cross-inoculation" group and that within such groups rhizobia from one plant would nodulate all other plants and vice versa. However, it soon became evident that these groups were not discrete and many reports of boundary jumping between groups are to be found in the literature.[164]

e. The Nodulation Process

Infection of root hairs, following the initial penetration or invagination of the hair cell primary wall, proceeds by the development of a tube whose inner surface is lined with

Table 10
CROSS INOCULATION GROUPS

Alfalfa group

Common name	Scientific name
Alfalfa	*Medicago sativa*
Button-clover	*M. Orbicularis*
California bur-clover	*M. denticulata*
Spotted bur-clover	*M. arabica*
Black medic	*M. lupulina*
Snail bur-clover	*M. scutellata*
Tubercle bur-clover	*M. tuberculata*
Little bur-clover	*M. minima*
Tifton bur-clover	*M. rigidula*
Yellow alfalfa	*M. falcata*
White sweet clover	*Melilotus alba*
Huban sweet clover	*M. alba annua*
Yellow sweet clover	*M. officinalis*
Bitter clover (sour clover)	*M. indica*
Fenugreek	*Trigonella foenumgraceum*

Clover group

Common name	Scientific name
Alsike clover	*Trifolium hybridum*
Crimson clover	*T. incarnatum*
Hop clover	*T. agrarium*
Small hop clover	*T. dubium*
Large hop clover	*T. procumbens*
Rabbit foot clover	*T. arvense*
Red clover	*T. pratense*
White clover	*T. repens*
Ladino clover	*T. repens* (giganteum)
Subclover	*T. subterraneum*
Strawberry clover	*T. fragiferum*
Berseem clover	*T. alexandrinum*
Cluster clover	*T. glomeratum*
Zigzag clover	*T. medium*
Ball clover	*T. nigrescens*
Persian clover	*T. resupinatum*
Carolina clover	*T. carolinianum*
Rose clover	*T. hirtum*
Buffalo clover	*T. reflexum*
Hungarian clover	*T. pannonicum*
Seaside clover	*T. wormkskjoldii*
Lappa clover	*T. lappaceum*
Bigflower clover	*T. michelianum*
Puff clover	*T. fucatum*

Pea and Vetch group

Common name	Scientific name
Field pea	*Pisum arvense*
Garden pea	*P. sativum*
Australian Winter Pea	*P. sativum* (var *arvense*)
Common vetch	*Vicia sativa*
Hairy or winter vetch	*V. villosa*
Horse or broad bean	*V. faba*
Narrow leaf vetch	*V. angustifolia*
Purple vetch	*V. atropurpurea*

Table 10 (continued)
CROSS INOCULATION GROUPS

Common name	Scientific name
Monantha vetch	*V. articulata*
Sweet pea	*Lathyrus odoratus*
Rough pea	*L. hirsutus*
Tangier pea	*L. tingitanus*
Flat pea	*L. sylvestris*
Lentil	*Lens culinaris (esculenta)*

Cowpea group

Common name	Scientific name
Cowpea	*Vigna sinensis*
Asparagus bean	*V. sesquipedalis*
Common lespedeza	*Lespedeza striata*
Korean lespedeza	*L. stipulacea*
Sericea lespedeza	*L. cuneata*
Slender bush clover	*L. virginica*
Striped crotalaria	*Crotalaria mucronata*
Sun crotalaria (sunnhemp)	*C. juncea*
Winged crotalaria	*C. sagittalis*
Florida beggarweed	*Desmodium tortuosum*
Tick trefoil	*D. illinoense*
Hoary tickclover	*D. canescens*
Kudzu	*Pueraria thunbergiana*
Alyce clover	*Alysicarpus ginalis*
(no common name)	*Erythrina indica*
Pigeon pea	*Cajanus cajan (indicus)*
Guar (cluster bean)	*Cyamopsis tetragonoloba*
Jackbean (horse bean)	*Canavalia ensiformis*
Groundnut (peanut)	*Arachis hopogaea*
Velvet bean	*Stizolobium deeringianum*
Lima bean	*Phaseolus lunatus (macrocarpus)*
Adzuki bean	*P. angularis*
Mat bean	*P. aconitifolius*
Mung bean	*P. aureus*
Tepary bean	*P. acutifolius* var *latifolius*
Acacia	*Acacia linifolia*
Kangaroo-horn	*A. armata*
Wild indigo	*Baptisia tinctoria*
Hairy indigo	*Indigofera hirsuta*
Partridge-pea	*Chamaecrista fasciculata*

Bean group

Common name	Scientific name
Garden bean, kidney bean, navy bean, pinto bean	*Phaseolus vulgaris*
Scarlet runner bean	*P. coccineus (multiflorus)*

Lupine group

Common name	Scientific name
Blue lupine	*Lupinus angustifolius*
Yellow lupine	*L. luteus*
White lupine	*L. albus*
Washington lupine	*L. polyphyllus*
Sundial	*L. perennis*
Texas bluebonnet	*L. subcarnosus*
Serradella	*Ornithopus sativus*

Table 10 (continued)
CROSS INOCULATION GROUPS

Common name	**Scientific name**
	Soybean group
All varieties of soybean	*Glycine max (soja max)*
Birdsfoot trefoil	*Lotus corniculatus*
Big trefoil	*L. uliginosus*
Foxtail dalea	*Dalea alopecuroides*
Black locust	*Robinia pseudoacacia*
Trailing wild bean	*Strophostyles helvola*
Hemp sesbania	*Sesbania exaltata*
Kura clover	*Trifolium ambiguum*
Sanfoin	*Onobrychis vulgaris (sativus)*
Crown vetch	*Coronilla varia*
Siberian pea shrub	*Caragana arborescens*
Garbanzo (chick pea)	*Cicer arietinum*
Lead plant	*Amorpha canescens*

cellulose. It contains rhizobia lying end to end, usually in polysaccharide matrix. The development of this tube, the infection thread has been documented by time lapse cinematography.[223] It can be regarded as an invagination of the root hairs and development of nodules.

The infection thread penetrates through and between the root cortex cells towards the stele. Actual nodule initiation, usually at the inner cortical tetraploid cells, takes place near diploid cells in the pericycle that has the potential of lateral root formation. A fully developed nodule consists of a nodule cortex, vascular system, meristematic and bacteroidal zone. The bacteroidal zone is located in the central portion of the nodule and is literally the bacterial tissue.

Once liberated from the infection thread into the root cytoplasm, the rhizobia assume a particular morphology and are termed as bacteroids. The bacteroids within the nodule are swollen and irregular frequently appearing in star, clubbed or branched shapes (Figure 3).[224] The principal feature of internal structure of normal bacteroids is a fairly even distribution of electron-dense material. The mature bacteroids are enclosed by two membranes in close juxtaposition (Figures 4 and 5).

The infection of a legume by *Rhizobium* is an intimate association which depends on their total specific mutual recognition for effective nitrogen fixation. Steps in establishing the symbiosis involve colonization of the rhizosphere, entrance (via the root hair or during the emergence or lateral roots) resulting in the formation of infection threads, commencement of a persisting nodule meristem, release of rhizobia from the infection thread, their multiplication within membrane envelopes of the nodule host cell, conversion to nodule bacteroids, the establishment and continuance of shared metabolism between plant and bacterium. This intimate association between *Rhizobium* and host requires all aspects of the relationship to be mutually acceptable for effective nitrogen fixation. In all associations, both the bacterial and plant partners represent a wide range of properties, the effectiveness of nitrogen fixation depends on the degree of compatibility between two partners. Thus, a *Rhizobium* effective on one host does not necessarily result in an effective association in a closely related host species. Nonnodulated plants probably fail during the early stages of infection while the commonly observed ineffective associations fail at a later stage. Usually a nodule is occupied by a single strain of *Rhizobium* but there are reports of dual representation by different strains of similar rhizobia.[225]

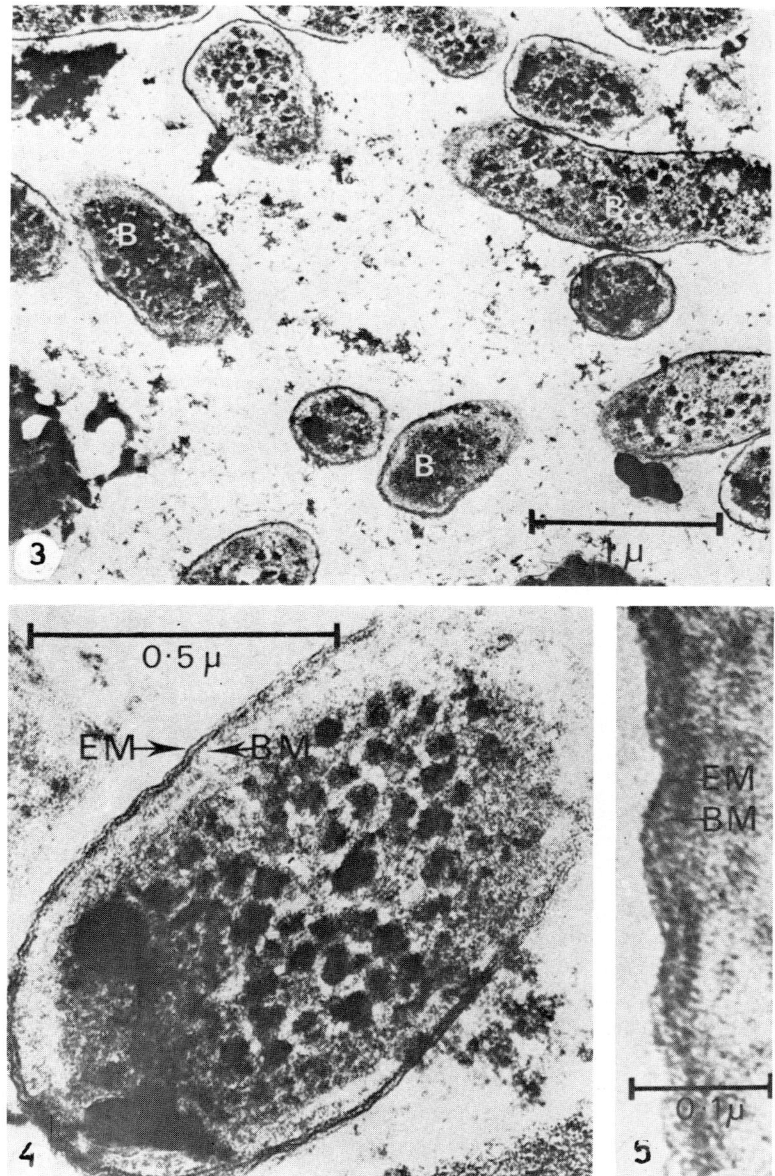

FIGURE 3. Section of clover root nodule showing normal bacteroids. FIGURE 4. Mature normal bacteroid showing double membrane and electron dense material in the cytoplasm. FIGURE 5. Enlargement of membrane in Figure 4. (From Fisher, D. J., Hayes, A. L., and Jones, C. A., *Ann. Appl. Biol.*, 90, 73-89, 1978. With permission.)

Much is not known about the molecular mechanisms controlling host specificity or the biochemical basis for the cellular recognition in legume-*Rhizobium* symbiosis. There are many stages during the complex process of nodule development when recognition phenomena may occur and the control of host specificity be exerted.[226] A survey of the literature suggests that host specificity is not absolute at any single stage but that it is expressed to a considerable extent at or before penetration of the host hair and formation of infection threads[227-230] when control of specificity must manifest itself at one or all of four stages.

1.   Root colonization
2.   Root hair adhesion
3.   Root hair curling
4.   Root hair penetration

During recent years Dazzo and colleagues have done a considerable amount of work on recognition and the determinants of host specificity in *Rhizobium trifolii* colonizing system. Dazzo and Hubbell[231] proposed a cross-bridging model of lectin involvement. They postulated that the clover root lectin, trifoliin A recognized unique cross reactive antigens thus forming a correct molecular interfacial structure to initiate the preferential and specific adsorption of the bacteria to host root-hair surface. Dazzo and co-workers have accumulated much experimental evidence to support their cross-bridging hypothesis.[232-234]

Rhizobium spp. are largely defined by host plant range within the leguminosae.[235] In homologous (nodule productive) combination of bacteria and plants, microscopic studies reveal that bacteria attach to plant cells, the root hairs of the host markedly curl and host cells are invaded by way of infection threads.[217,236,237] In heterologous (non-nodulating) combination of bacteria and host, host root hairs may show partial deformation but no curling.[230] Genetic studies of nodulation by fast growing *Rhizobium* strains have demonstrated a series of loci required for nodulation (nod genes) which are linked to nitrogenase (nif) genes on very long symbiotic plasmids.[238-243] Mutations in these nod genes result in the failure of nodule development at early stages. In *Rhizobium meliloti*, nod mutants have been isolated which fail to curl root hairs.[244]

In crosses between *Rhizobium leguminosarum* and *Rhizobium trifolii* host range selectivity is co-transferred with other nodulation loci[239,245,246] suggesting that nod and host-range genes are either identical or closely linked to these species. Recently Fisher et al.[243] have reported several nodulation genes in *Rhizobium meliloti* replaceable by cloned nod gene DNA fragment of *Rhizobium trifolii* and that *Rhizobium meliloti* clones likewise complement to *Rhizobium trifolii* nod genes.

f. Rhizobium Inoculation

The inoculation of legumes with root-nodule bacteria (*Rhizobium* spp.) is one of the few cases of man's successful exploitation of microorganisms for agricultural purposes. He has been able to select and multiply suitable bacteria and use them as bacteria fertilizers. There is little argument about need for inoculation in most of the world's agricultural soils. The understanding of the nature of the inoculation of legumes with rhizobia is essential because invariably seed inoculation with rhizobia is followed by fungicide dressing to protect the seed from many fungal diseases. However, fungicides and rhizobia are never compatible and there are always some toxic effects on the rhizobia applied to seeds. Therefore, understanding of the nature of inocula will be of much significance in applying the fungicides to seeds.

Earlier in the beginning of the century, soil transfer was the recommended method of legume inoculation. Undesirable features inherent in this onerous method were the high cost of transporting the soil any distance and the danger of introducing weed seeds, insect pests, nematodes, and organisms effecting plants and animals. Accordingly the method was soon replaced by artificial inoculation of seed first by the use of cultures of nodule bacteria grown on agar medium and later by peat of humus cultures.[225] The first commercial inoculants were produced on gelatin and later on agar nutrient media. One of the techniques recommended for supplying agar-based inoculant was application of suspension directly to the seed. Commercial liquid inoculants (by growing bacteriological material in jars or in pots in a glasshouse in liquid medium) are prepared by soaking the seed in a culture suspension before sowing, by applying it to the newly sown seeds, or to the roots of the young seedlings. In addition, solid-based inoculants now from vast majority of culture were marketed.

2. Ecology of Legumes
a. Distribution and Population Structure

The legumes have achieved cosmopolitan distribution and importance. Representatives of the family are found in all continents, from the hot, humid Asian lowland tropics to the cold, barren Siberian steppes; from sea level to montane meadows, and from the semidesert savanna of Africa to the cool altiplans of South America. Edaphic conditions range from dry, coarse textured, stony hillside sites in the Middle East to the fertile riverbank sites of Burma. Important legumes may be found growing from the equator to the Arctic Circle, and at least two useful species (*Medicago falcata* and *Trifolium lupinaster*) survive winter temperatures as cold as $-67.8°C$ on the tundra at Verkhoyansk, Siberia (latitude 68°N), where the subsoil is frozen permanently.[247]

Only the grasses exploit greater ecological diversity than the legumes, although they are perhaps less diverse morphologically. This is because legumes share certain basic forms and structure common to grasses and the economic legumes possess adaptive versatility in their morphological patterns, anatomical variations, and physiological behavior that has been developed through the processes of natural selection and domestication, and is maintained by species-specific breeding systems. It is this versatility that renders a relatively small number of legumes uniquely fitted to the environments in which they occur, and to the useful functions they serve. Plant population geneticists have found it appropriate to recognize three categories of species as regards to their structure.

i. Predominantly Self-Fertilized

Such legumes exhibit extensive variability, both morphologically and physiologically. However, in certain other critical respects, they display a very limited range of variation and this severely restricts them to certain ecological zones. *Phaseolus vulgaris* is the typical example of this type of legume. Most plants are extremely homozygous, outcrossing is rare (less than 3%) except in the usual situation where wild pollinators abound. In domesticated *Phaseolus vulgaris*, variability is between cultivars and only minor variations occur within cultivars.

Intregression from wild forms is essentially nil since present habitats of domestic and wild beans do not overlap. Vigorous hybrids can be produced from crosses between diverse types and there is evidence of transgressive segregation in some crosses. There is little information on the extent of free recombination. Certain cultivar crosses generate almost a depauperate array of phenotypes, with very few segregants equal in overall merit to either parent. This suggests that genotypic balance is an important cohesive force which influences population fitness.

ii. Normally Cross-Fertilized

Medicago sativa and *Medicago falcata* require cross-fertilization for abundant seed set and for maintenance of normal population vigour. Such species are often partially or even obligatorily self-incompatible, or they may be self-fed for only a limited number of generations. Inbreeding is nearly always accompanied by decline of vigor and fertility.

Cross-incompatibility systems in some herbaceous perennial legume species restrict matings among close relatives, and this further reduces inbreeding and genetic homozygosity. The modal-class plant is expected to be extremely heterozygous for most loci, similar in phenotype but different in genotype from other modal class plants.

A species such as *Medicago falcata* and even some forms of *Medicago sativa*, is more likely to survive for a prolonged period in the wild than the domesticated common bean, pea, or soybean. Since the latter species are truly domesticated they probably have lost in the process the special characteristics essential to survival in natural communities, whereas, *Medicago falcata* has hardly been changed in any significant way from its natural ancestors.

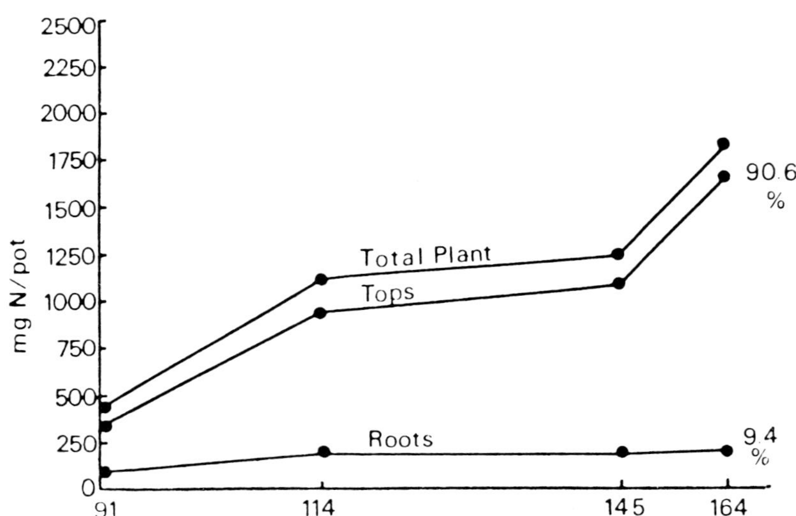

FIGURE 6.   Total N accumulated by crimson clover at different stages of growth. (From Giddens, J., *Univ. Ga. Res. Bull.*, 327, 1-37, 1985. With permission.)

However, there is an alternative and more speculative, reason which can be evoked. The falcate native to the barren steppe of Russia, primarily had to adapt itself for survival against the cold and drought of a rigorous and relatively unchanging physical environment — by combination of genetically conditioned perenniality, the cold resistance, it has been able to develop, and by its breeding system, to maintain or even improve its capability to survive. In comparison, the bean, native to the temperate tropics, had to contest for survival in an environment characterized by dominant biological forces, such as other strongly competitive plants and complexes of diseases and insects which are also capable of change against which the natural breeding system of the intervention has proved thus far been unable to cope.

iii. Often Cross-Fertilized

In the evolutionary sense, species of this class (e.g., *Cajanus cajan*) share the benefits of two breeding systems. A moderate amount (1 to 40%) of out-crossing continually or intermittently generates hybrids which will undergo subsequent self-fertilization to produce a segregating array of recombinants, many of which may become extremely homozygous before being involved in cross-fertilization again.

b. N₂-Fixation in Legumes

The amount of N_2-fixed by legumes can be considerable.[248] For this reason legumes have also been employed as green manure. The patterns of nitrogen fixation and the amount of nitrogen fixed is dependent on many factors. Legume in association with rhizobia generally continue fixing nitrogen through their life span. Nelson et al.[249] showed that acetylene reduction for soybeans continued at a much higher rate until late maturity when plants were grown in nondestructive system where plants roots were not cut-off for making the assay. Giddens[6] investigated the patterns and amounts of nitrogen fixed by different legumes and Figures 6 to 9 illustrate that a major part of N for these legumes occur in the above ground portion. The roots of crimson clover and arrowleaf clover contain only a minor part of the N, while clover and alfalfa roots contain a much larger part. Roots represented 9.4% of the N for crimson clover, 15.4% for arrowleaf clover, 30.3% for white clover, and 48.4% for alfalfa. It is interesting to note that where these crops are closely grazed or harvested for forage, only white clover and alfalfa contribute appreciable amounts of N to the soil. The

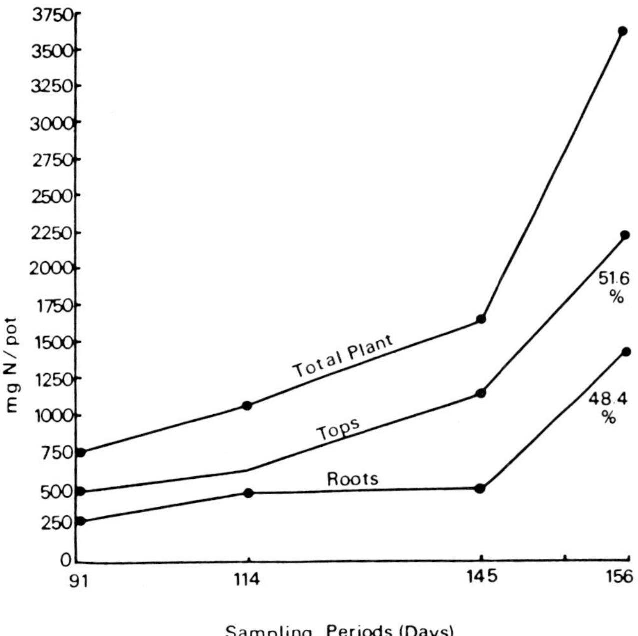

FIGURE 7.   Total N accumulated by alfalfa at different stages of growth. (From Giddens, J., *Univ. Ga. Res. Bull.*, 327, 1-37, 1985. With permission.)

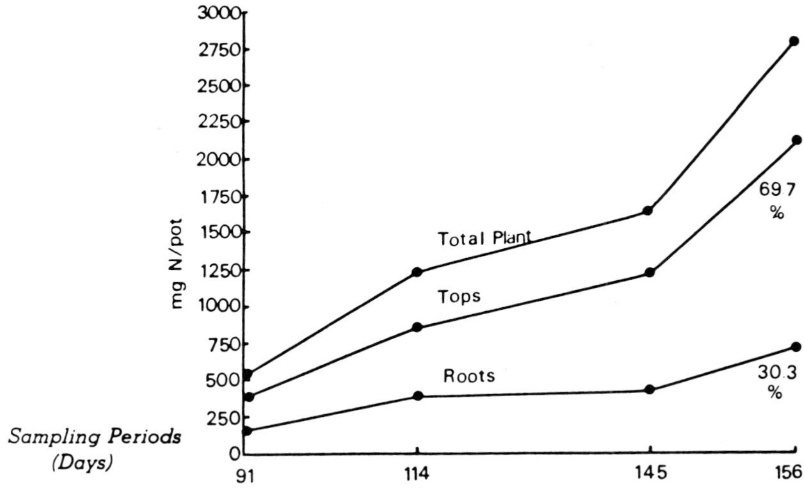

FIGURE 8.   Total N accumulated by white clover at different stages of growth. (From Giddens, J., *Univ. Ga. Res. Bull.*, 327, 1-37, 1985. With permission.)

fixation of nitrogen and its distribution in legumes is also considerably different.[6] Most of the total N accumulated in the leaves and petioles of soybeans until the midpod-fill stage (Figure 10), at which time all plant parts lost N to the pods. By late maturity 92% of the total N had accumulated in the pods of this cultivar. The peak period of C_2H_2 reduction for soybeans is at the midpod-fill stage (Figure 11). Thus, it is evident that the prior knowledge of the patterns of N_2-fixation and contribution of each legume to N economy and distribution of N contents is essential for studying the effects of pesticides on legume partner of the symbiotic association.

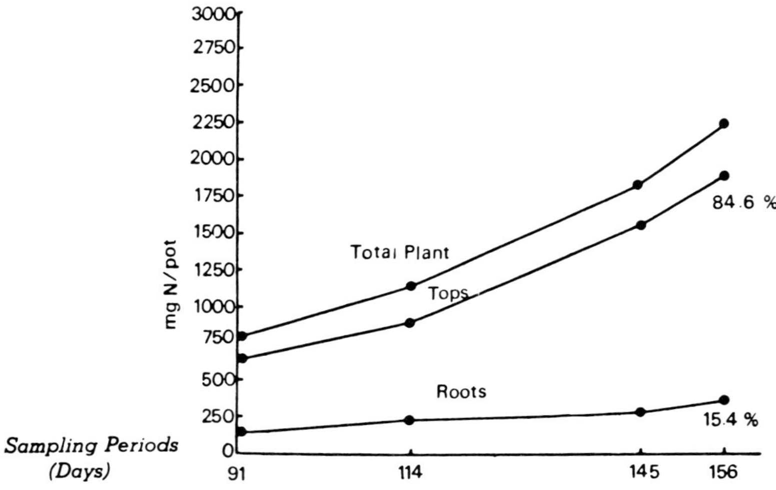

FIGURE 9.   Total N accumulated by arrowleaf at different stages of growth. (From Giddens, J., *Univ. Ga. Res. Bull.*, 327, 1-37, 1985. With permission.)

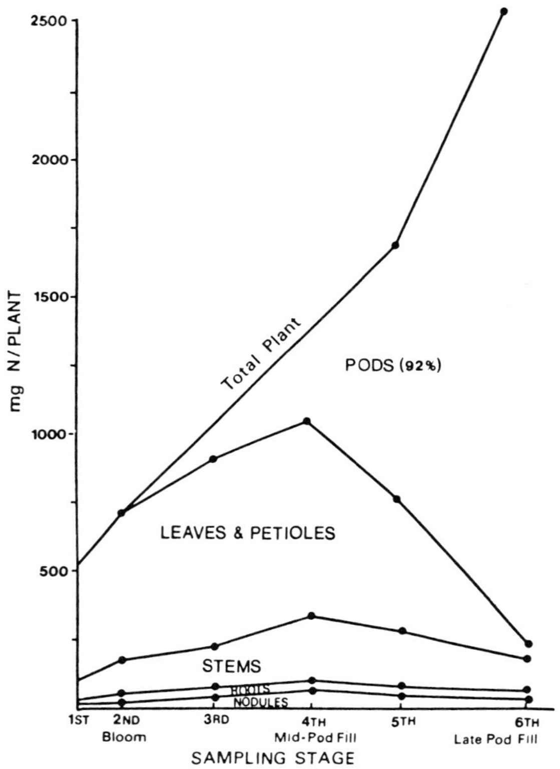

FIGURE 10.   Total N accumulated by soybeans at different stages of growth. (From Giddens, J., *Univ. Ga. Res. Bull.*, 327, 1-37, 1985. With permission.)

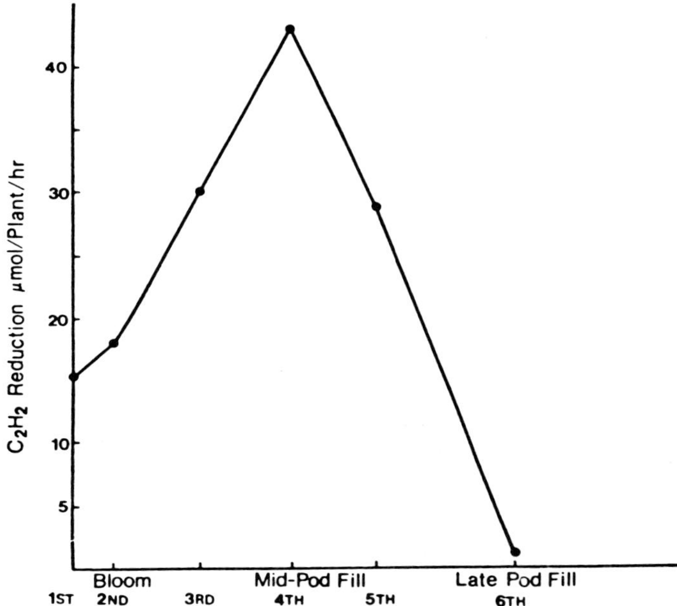

FIGURE 11. Nitrogen fixation (C_2H_2 reduction) by soybeans at different stages of growth. (From Giddens, J., *Univ. Ga. Res. Bull.*, 327, 1-37, 1985. With permission.)

c. Environmental Influence

Periodic environmental stresses of varying duration and kind, are a common feature during the growth of any legume. In order to understand the effect of pesticides on legume partner it is essential to understand the effect of these factors. In general the amount of nitrogen fixed by the legumes is influenced by the light, moisture, lowered oxygen concentration, adverse soil pH, and nutrient deficiencies. In addition, the number of rhizobia (to effect prompt, adequate nodulation) also influence the nitrogen fixation.

The normal practice of reviewers dealing with the effect of environment on legume symbiosis has been to deal with the various factors known to influence nodulation, plant growth, nitrogen fixation, and yield. This approach has its merits as only a comprehensive knowledge of this environment can provide the basis for studying pesticide legume interactions. The influence of environmental factors has been dealt in several reviews.[250-254]

Apart from its special ability to fix atmospheric nitrogen, the nodulated legume probably functions in a basically similar way to any other green higher plant, so that dissimilarities between legume and nonlegume are likely to be relative and to stem largely from the nutritional and functional interdependencies of the nodule plant system rather than from any inherent properties of the host plant.

i. The Carbon Economy of the Host Plant and Nodules

Nodules in general maintain meager reserves of a readily utilizable carbohydrate relative to their requirements for nitrogen fixation so that they probably rely heavily for the growth and functioning on photosynthetic products, currently translocated from the leaves, or on reserves of carbohydrates mobilized from other regions of the plant.[255-257] There are many reports in which reduced photosynthesis, by reducing light or defoliation causes drastic or immediate reduction in fixation.[254,258-260] Improvement of photosynthetic capacity either by increasing day length or light intensity, or by increasing the ambient concentration of CO_2 around a photosynthesizing shoot[261] or by grafting onto the plant a second shoot system,[262] stimulates nitrogen fixation. Further removal of root apices or fruits[263,264] can cause increase in nitrogen fixation. It is difficult to explain with reason the changes in nitrogen fixation but all these factors are linked with the physiology of nitrogen fixation by the legume.[265,266]

The supply of carbohydrates from the plants to the nodule is also dependent upon the diurnal variation and nodules detached from the plants at night fix nitrogen poorly compared with those detached at day time.[259,267-270] This is because the levels of carbohydrates fall during night time because of no photosynthesis whereas the reverse happens during day time.[257] However, periods of darkness is the time for high nodule activity provided that sufficient carbohydrates have built up in the nodule during the previous photoperiod and the night temperature is optimum for efficient use of carbohydrate for fixation.[271]

ii. Nitrogen Economy of the Host Plant

The nitrogen fixed by nodules proceeds in an upward direction to the shoot through xylem,[253] thus the entire amount of nitrogen fixed is distributed among various plant parts[272] and is incorporated into amino acids which are finally secreted into the xylem.[256,273] The nitrogenous products exported from nodules in the xylem primarily consist of nitrogen rich molecules such as amides or ureides,[253,274] and on arrival in the shoot these amino compounds either serve as substrate for protein synthesis or get converted to soluble nitrogen to be utilized for seed nutrition.[275,276] The nitrogen received by the shoots from nodules finally returns to the root in the form of amino compounds in the phloem along with amino compounds generated in the photosynthesis.[253] Thus, the nitrogen cycling from root to shoots and vice versa also controls the process of nodulation.[274,277,278]

Nodulation and fixation of nitrogen is also dependent on the nitrate supply in the rooting medium with higher quantities of nitrate nitrogen suppressing the nitrogenase activity.[278] There are convincing evidences that small supplements of combined nitrogen in legumes can stimulate nitrogen fixation if correctly timed in their application.[251,254,279] This can be exploited to obtain higher yield of legume plants by providing fertilizer nitrogen during early establishment of the seedlings[253] or during fruit filling when there is a peek demand for nitrogen synthesis. Sufficient N-fertilizer (50 kg/ha of N fertilizer) was required to produce forage and harvestable beans equal to the inoculated beans in Georgia soil (Table 11).[6] The percentage N in the seeds of soybean treated with the highest N application rate (374 kg/ha) was equal to the nodulated plants and 250 kg/ha N was required on nonnodulated beans to equal the bean yield of the nodulated cultivar.

iii. The Water Economy of Host Plant and Nodules

The export of fixed water is essential as it maintains the turgidity of the tissue. Nodular activity is acutely suppressed when whole plants or detached nodules are subjected to water stress.[280-283] The requirements of water vary with the type of legume and the physiological stage of the plant.[253] Loss of fixation activity on desiccation occurs in nodules with determinate growth[280] or indeterminate growth.[283,284] The effects of water stress are reversible provided that water loss from nodule does not account more than 20% of its maximum fresh weight.[281] However, irreversible structural damage occurs with severe draught damage which results in the collapse of cells in the nodule cortex.[281]

Moisture not only limits the N_2-fixation in soybeans but it also limits the supply of nitrogen from the soil to nonleguminous plants. For instance Giddens[6] found that with the moisture stress that occurred in Georgia soil, moisture became a limiting factor before N became limiting to corn yield grown in soils with legume as green manure (Table 12). The corn yield values reported in Table 13 indicate that where part of the moisture stress was alleviated by one or two supplement irrigations (during 1980 and 1982) high corn yield was found even with a smaller supply of nitrogen from previous grown soybean crop as green manure.

Table 11
RESPONSE OF NODULATING AND NONNODULATING LEE CULTIVAR SOYBEANS TO NITROGEN FERTILIZER AND RESPONSE OF THE NODULATING CULTIVAR TO TWO RATES OF INOCULATION WITH STRAIN 110(USDA) *RHIZOBIUM*

| | Forage yields (kg/ha)[a] | | N (%) | | | | Bean Yield (Kg/ha) | |
| | | | Forage | | Seed | | | |
Treatments	1976	1977	1976	1977	1976	1977	1976	1977
Nl not inoculated	1822	2239	2.88[a,b]	2.91[a]	6.25[a]	6.27[a]	1780[a,b]	2336[a]
N2 inoculated, low rate[b]	2184	2350	2.77[a,b,c]	2.91[a]	6.22[a]	6.44[a]	1897[a]	2321[a]
N3 inoculated, high rate[c]	2210	2339	2.79[a,b,c]	2.94[a]	6.37[a]	6.34[a]	1824[a]	2273[a]
N4 inoculated, high rate + 250 kg N	2037	2419	3.02[a]	2.86[a]	6.36[a]	6.38[a]	1835[a]	2262[a]
NN1 O N	1814	2101	2.31[d]	1.19[c]	4.65[b]	3.86[d]	981[c]	984[c]
NN2 50 kg/ha N[d]	2246	2401	2.38[c,d]	2.30[b]	5.32[b]	5.21[c]	1556[a,b]	1885[b]
NN3 125 kg/ha N	2107	2276	2.52[b,c,d]	2.76[a]	4.96[b]	6.00[b]	1256[b,c]	2363[a]
NN4 250 kg/ha N	2050	2073	2.67[a,b,c,d]	2.89[a]	5.23[b]	6.28[a]	1287[b,c]	2197[a]
	NS	NS						

Note: N represents nodulating and NN represents nonnodulating Lee soybeans. Values in each column followed by the same letter are not significantly different at the 5% level.

[a]   Weight of forage at bloom stage.
[b]   Inoculated with 10^4 rhizobia per centimeter of row.
[c]   Inoculated with 10^8 rhizobia per centimeter of row.
[d]   In 1977 N rates changed to N_2-125, N_3-250, and N_4-375 kg/ha.

From Giddens, J., *Univ. Ga. Res. Bull.*, 327, 1-37, 1985. With permission.

Table 12
EFFECT OF SOYBEANS AND N RATES ON CORN GRAIN YIELD, 1976 TO 1977

| | 1976 (Kg/ha) | | | 1977 (Kg/ha) | | |
Treatment	O N	60 N	120 N	O N	60 N	120 N
Fallow; no soybeans	4378	4284	4999	3456[a,b]	3299[a,b]	4089[a]
Soybeans; harvest pods	3933	4723	4171	3562[a]	3657[a]	4133[a]
Soybeans; harvest above ground	4039	4848	4754	2465[b]	3738[a]	4077[a]
	NS	NS	NS			

Note: Soybean yield in 1975 was 2016 kg/ha (121 kg N). Soybean yield in 1976 was 1169 kg/ha (71 kg N). Values in rows or columns followed by the same letter are not significantly different at the 5% level.

From Giddens, J, *Univ. Ga. Res. Bull.*, 327, 1-37, 1985. With permission.

iv. The Gasous Environment of the Nodulated Root and Symbiotic Functioning

Legumes are generally intolerant of waterlogged or poorly aerated soils. Aerated nodules, fix maximum nitrogen,[69] and lowering oxygen level in the aerating system causes a great suppression of fixation.[285] In the waterlogged environment, a decline in nodular fixation down a root can be observed.[253] In sand soil the nitrogen fixation if other conditions are

Table 13
EFFECT OF SOYBEANS AND N RATES ON CORN GRAIN YIELD, 1978, 1980, AND 1982

Treatments	1978 (Kg/ha)			1980 (Kg/ha)			1982 (Kg/ha)		
	O N	60 N	120 N	O N	60 N	120 N	O N	60 N	120 N
Fallow; no soybeans	4685	4754	4999	5105[a,b,c]	4748[b,c]	4905[a,b,c]	4955[c]	7608[a,b]	8862[a]
Soybeans; harvest pods	3657	4961	5018	4622[c]	5864[a,b,c]	6548[a]	3989[c]	6498[b]	7853[a,b]
Soybeans; harvest nothing	4014	4246	3939	6354[a,b]	6372[a,b]	6535[a]	6604[b]	7715[a,b]	7294[a]
	NS	NS	NS						

Soybean yield 2509 kg/ha, 161 kg N   Soybean yield 1857 kg/ha 120 kg N   Soybeans yield 1767 kg/h 106 kg N

Note: Values in each row or column followed by the same letter are not significantly different at the 5% level.

From Giddens, J., *Univ. Ga. Res. Bull.*, 327, 1-37, 1985. With permission.

normal is maximum.[284,286] Respiration and nitrogen fixation studies on detached nodules exposed to different oxygen tension show that efficiency of carbohydrate consumption in fixation is the highest at near atmospheric levels of oxygen and decreases with lowering of oxygen supply.[287] Oxygen tension affects the symbiotic association and plant metabolism. The latter effect will also finally decrease the nitrogen fixing ability.[288,289] On the contrary, acetylene is injurious to nitrogen fixation process[290] and concentrations of carbon dioxide in excess of 1% are stimulatory in certain legumes.[290]

v. Temperature

Legumes respond to lower temperatures by producing a greater volume of nodule tissue than is otherwise anticipated and that this may be regarded as a form of compensation against lower nitrogen-fixing activity per unit of tissue.[250] In *Trifolium subterraneum* an increase in dry weight and nitrogen content was found in the nodules when the plants were grown at 8°C relative to those of 15 and 15°C relative to those of 22°C.[291,292] Similarly, lower temperature has been reported to increase leghemoglobin in individual nodules.[293] Lupins and soybeans produced a higher proportion of their total dry weight and nitrogen in nodule tissue grown at suboptimal than at optimal temperatures and at lower temperature legumes showed greater capacity for nitrogen fixation than the nodules from the plants grown at higher temperatures. Responses to higher temperature conditions depend upon the strain of rhizobia and the host plant.[272,292,294] Temperature effects are mediated through fundamental influences on the rates of water movement and the uptake and transport of mineral ions; on the rates of respiration, photosynthesis, transpiration and translocation, on membrane integrity (phospholipid composition), and ultimately cytoplasmic function. Outward manifestations include differences between species in the critical temperatures for seed germination, vegetative growth, flowering, seed setting, and in the case of perennial species, for storage and dormacy.

Thus, certain legumes are recognized as distinctly tropical (e.g., cowpeas, pigeon peas, the mungs orgrams, and groundnuts), the others are better adapted to subtropical or to temperate climate (e.g., the *Phaseolus* beans, chick peas, faba beans, field peas, soybeans, and lentils) whereas the small seeded forage legumes are primarily temperate zone species.

The initiation and expansion of flower buds into open flowers also depends upon temperature, although in field studies it is generally not possible to separate these effects from those of daylength. One system which allows this is to install flood lamps in the field and to plant at varying distances from the light source in order to simulate a range of daylengths.

C. *Azospirillum*

Nitrogen fixation in legumes through the *Rhizobium*-legume symbiosis is well known. But cereal crops which provide the bulk of our staple food are largely dependent on chemical fertilizers. Until recently, these plants were believed to have no nitrogen fixation potential. However, a new nitrogen-fixing *Azospirillum* was recently isolated from the roots of crops like wheat, maize, and sorghum. Initially these organisms were observed in the mucilaginous sheath on the root surface. But later studies revealed that this bacterium can enter inside the roots and colonize the cortical cells of many gramineae plants.

The basic difference between this bacterium and *Rhizobium* is that it does not form any nodules on the host root nor is it dependent on the host obligatory. Instead it forms a loose association with roots utilizing the root exudates. This new type of association is termed as associative symbiosis or semisymbiosis. Unlike *Rhizobium*, *Azospirillum* can fix nitrogen in a culture medium as efficiently when provided with necessary nutrients. It is an aerobic organism but can fix nitrogen only under microaerophilic conditions where the partial pressure of oxygen (pO_2) is around 0.002 atm. Hence it is normally grown in semisolid media. The genus *Azospirillum* is divided into two species; *Azospirillum lipoferm* and *Azospirillum brasilense* based on DNA homology and other physiological characters.

The main reason why this organism is attracting the attention of the scientists is its high nitrogen fixation ability compared to other free living bacteria like *Azotobacter*. Another important factor is its close association with cereal plants which need high amounts of chemical fertilizers and the possibility of meeting some of these requirements through inoculation with this organism. The nitrogenase in *Azospirillum* fixes N_2 only under microaerophilic conditions, whereas in case of *Rhizobium* the nitrogenase is protected by leghemoglobin.

Azospirillum is abundant in tropical soils where it may have particular potentialities for enhancing plant growth. It occurs also in temperate soils.[295-297] There is currently much debate about the beneficial effects of free living or root associated *Azospirillum*.[298,299] It may augment plant growth by providing products of N_2-fixation, either by a direct bacterium-plant transport of fixed nitrogen or by a slow transfer due to a gradual death of bacteria and their mineralization.[300-303] The bacterium may also increase plant productivity by producing phytochromes[304] and/or by enhancing mineral uptake in general.[305]

However, the issue is controversial as the situation is complicated by the reports that *Azospirillum* can perform denitrification.[306] Under anaerobic conditions in liquid cultures, *Azospirillum* utilizes nitrate, nitrite, or nitrous oxide or molecular nitrogen depending upon the assay condition and strain used.[307-311] Thus, the opposite from the desired may happen by a plant-*Azospirillum* association under unfavorable conditions.[303]

D. Blue-Green Algae

Recognition of blue-green algae (Cynobacteria) as nitrogen fixer dates back to 1889 when Frank noticed that soil containing blue-green algae-fixed nitrogen. The same year Prantil also noticed that impure cultures of *Nostoc*-fixed nitrogen. This was subsequently confirmed in 1928 by Drew who reported that *Anabaena variabilis*, *Anabaena* sp., and *Nostoc punctiforme* grew well in a pure culture with a medium free of combined nitrogen.[312] More extensive work on the fixation of nitrogen by blue-green algae, however, became possible by the developments of such techniques as acetylene reduction and ^{15}N assay. The list which now includes many blue-green algae known to fix nitrogen is given in Tables 14 and 15.

There are three main groups of blue-green algae: unicellular, filamentous heterocystous, and filamentous nonheterocystous. In all these organisms nitrogenase is responsible for the fixation of nitrogen. The nature of this enzyme is similar in all three groups of algae and the main differences appear to be in the localization of this enzyme and the ways in which it is protected from inactivation by oxygen.

The cells of blue-green algae are enclosed by a rigid, multilayered wall with an inner peptidoglycan layer. The wall may in turn be surrounded by a gelatinous or fibrous sheath. Most cynobacteria are mobile at some stage of the development; mobility is always of a gliding type, dependent on surface contact. The cytoplasmic region is traversed by an exclusive system of paired photosynthetic lammellae (thalakoids); the outer surface bears characteristic granules (phycobilisomes) composed of aggregates of the phycobiliprotein pigments.

Cynobacteria are either unicellular or consist of simple or branched colonies of cells. Reproduction of unicellular forms may occur by binary fission, or by aerial release of apical cells from a sessile individual. Filamentous forms grow by repeated intercalary cell divisions and reproduce either by random fragmentation of the filament or by terminal release of short motile chains of cells called hormogonia. Certain filament forms can produce specialized cells known as akinetes and heterocysts. Akinetes are larger than the vegetative cells in the filament and represent a resting stage, which germinates with a release of hormogonium. Heterocysts are nonproductive cells, distinguishable from the adjoining vegetative cells by the presence of refractile polar granules and of thick outer wall. They are believed to be physiologically specialized cells that serve as sites of nitrogen fixation.

Table 14
HETEROCYSTOUS BLUE - GREEN ALGAE
REPORTED TO FIX NITROGEN IN PURE
CULTURES

Anabaena ambigua	*Fischerella major*
A. azollae	*F. muscicola*
A. cycadea	*F.*
A. cylindrica	*Hapalosiphon fontinalis*
A. fertilissima	
A. flos-aquae	*Mastigocladus laminosus*
A. gelatinosa	
A. humicola	*Nostoc calcicola*
A. levanderii	*N. commune*
A. naviculoides	*N. cycadeae*
A. variabilis	*N. entophytum*
	N. muscorum
Anabaena sp.	*N. paludosum*
	N. punctiforme
Anabaenopsis circularis	*N. sphaericum*
Anabaenopsis sp.	
	Scytonema arcangelii
Aulosira fertilissima	*S. hofmanni*
Calothrix brevissima	*Stigonema dendroideum*
C. elenkinii	
C. parietina	*Tolypothrix tenuis*
C. scopulorum	*Westiellopsis prolifica*
Chlorogloea fritschii	
Cylindrospermum gorakhpurense	
C. licheniforme	
C. majus	
C. spgaerica	

Table 15
NONHETEROCYSTOUS BLUE-
GREEN ALGAE REPORTED TO
FIX NITROGEN IN PURE
CULTURES

Gloeocapsa 795	*Phormidium sp.*
Gloeocapsa 6501	*Plectonema boryanum*
Lyngbya 6409	*Plectonema 6306*
Oscillatoria 6407	*Plectonema 6402*
Oscillatoria 642	*Rhaphidiopsis indica*
Oscillatoria 6506	*Trichodesmium (impure)*
Oscillatoria 6602	

The characteristic photopigments include chlorophyll a as the only chlorophyllous pigment and phycobiliproteins (allophycocyanin and sometimes hypcoerithrin). The cellular absorption spectrum has a peak at approximately 665 nm attributable to chlorophyll a and broad absorption band with one or more peaks, attributable to phycobiliproteins between 560 to 630 nm.

1. Habitats of Nitrogen-Fixing Blue-Green Algae

Blue-green algae are widespread and abundant in a variety of marine, freshwater, and terrestrial habitats. The principal restrictions on their occurrence are that they usually require light for growth, that they do not grow well under acid conditions, and they (for reasons that are not clear) are not abundant in the open waters of the colder seas. Otherwise the resistance to extremes of temperature, desiccation, and the independence to supplies of combined nitrogen and vitamins exhibited by many species, enable them to occupy an extraordinarily wide range of habitats.

In freshwater, particularly in neutral or alkaline eutrophic lakes, blue-green algae are especially conspicuous as components of the plankton. Their abundance often shows a correlation with the concentration of dissolved organic matter, a circumstance that is difficult to understand, since they do not require endogenous supply of vitamins and are not heterotrophic.[313] It has also been observed that they show maximum growth when concentrations of phosphate and nitrate in the water are at their lowest. Nitrogen fixation may sometimes, but not always be the reason for vigorous growth when combined nitrogen is scarce. It appears likely, however, that with nonnitrogen-fixing species and in cases of phosphorus deficiency the algae generally grow at the expense of the intracellular reserves of phosphate and nitrogen which they can accumulate.[313] In temperate lakes, maximum of blue-green algae tend to occur in late summer and autumn but tropical waters, especially those organically polluted, may have a bloom of these algae throughout the year. In certain tropical lakes the planktonic blue-green algae form the main food of large flocks of flamingoes. When the water column is thermally stratified, planktonic blue-green algae, such as *Oscillatoria subscens,* form sharply defined maxima at particular depths. This is accomplished by means of their gas vacuoles.[313-315] In deep water where light intensity is low and growth is slow, gas buoyancy is increased and the filaments rise in the water. Near the surface where light intensity is high, gas vacuoles may be diluted out by rapid growth, so that buoyancy is lost and the filaments sink. In addition, if sugars or other low-molecular weight products of photosynthesis accumulate, the resulting increase in turgor pressure of the cell may collapse gas vacuoles. This mechanism will operate to the least extent when the supply of mineral nutrients is ample and the products of photosynthesis are taken up in growth. Hence, as is observed in nature, the greater the concentration of nitrate and phosphate, the higher is the number of blue-green algae. In shallow, turbulent waters, gas vacuolation is determined by the average light intensity reaching the cells. During a sudden calm the whole population (especially if it consists of a species producing large colonies, such as *Anabaena flosaquae*) may rise to the surface to form the familar bloom of blue-green algae. The concentration of cells on the surface may be such that carbon dioxide is depleted locally, so that photosynthetis and the gas vacuole collapse mechanism cannot operate. Caught in this way at the surface, the population eventually decays. The ensuing unpleasantness renders the water unfit for drinking.

Benthic blue-green algae are less obvious in lakes, but if the water is shallow their biomass may be greater than that of planktonic species. In frigid antarctic lakes, an orange-brown benthic felt of *Phormidium* species is often the most conspicuous vegetation present. In temperate waters, a vertical zonation which shows seasonal variation has been recorded. Species of *Rivularia, Tolypothrix,* and *Scytonema* may contribute substantially to the nitrogen budget of the water by their nitrogen fixation. Both in lakes and running waters blue-green algae are often associated with calcium carbonate deposition, but it appears that this association arises because the habitat favors certain species of this group and these may play a mechanical role in the aggregation of the tufa rather than that the algae play an active part in precipitating calcium carbonate.

The bright coloration of hot-spring areas in all parts of the world is largely due to blue-green algae and carotenoid containing flexibacteria.[314] Blue-green algae are characteristic of alkaline and travartine-depositing springs. Because of the temperature gradients in such areas different organisms occupy different zones. *Synechococcus elongatus* and *Synechococcus lividus* are found in water of temperature up to 72 to 75°C; bacteria can grow at higher temperature than this. The total algal biomass is usually greatest in the temperature range between 50 to 56°C, in which *Mastigocladus laminosus, Phormidium laminosum*, and *Synechococcus* species characteristically predominate.[104] Below 45°C, *Calothrix* species become abundant and diatoms appear. Apart from the zonation downstream and transversely across the stream, there is also a critical zonation in the algal mats, with blue-green algae on the surface, flexibacteria beneath the bacteria other than flexibacteria below. The filamentous blue-green algae may move by gliding to another position if temperature becomes suboptimal. Rather high biomass encountered above 45°C has been ascribed to the absence of herbivores above this temperature.

A semiaquatic habitat that is particularly favorable for the growth of blue-green algae is afforded by rice fields.[315-317] These are submerged under about 10 cm of water for 60 to 90 days during the rice growth and the high temperature and the ample nutrient supply favor algal growth. If the soil is neutral or alkaline species such as *Anabaena, Tolypothrix*, and *Aulosira fertilissima* predominate. When the fields dry out at harvesting, the growth remains as a paper film on the soil, and the resistance of the algae to desiccation ensures that an inoculum survives for the next season. Among these algae many are nitrogen-fixing species, and these make a significant contribution to soil fertility. Estimates of the combined nitrogen added to the soil range from 1 to 50 kg/h/year, especially in countries such as India where only a small amount of fertilizer is put on the fields. Improving conditions for blue-green algae by applying lime, phosphate, and traces of molybdenum (which is specially required for nitrogen fixation), and ensuring that a vigorous nitrogen-fixing strain is present by mass inoculation have been shown in field trials to increase rice yields appreciably.

Now that the sensitive and convenient acetylene reduction technique has become available for the field determination of rates of nitrogen fixation, data are being accumulated that show that nitrogen fixation by blue-green algae in soil is much more extensive than was at one time supposed. Results from Sweden show that nitrogen fixation rate by blue-green algae is up to 51 kg N/h/year.[318] Apart from this, blue-green algae may be important in binding the soil surface and reducing erosion. Blue-green algae are also abundant on rocks (especially of calcareous nature), buildings (especially in humid tropics), tree trunks, and leaves.

Blue-green algae are not abundant in the plankton of temperate and polar seas. However, in tropical waters *Trichodesmium* species are often extraordinarily abundant.[315] Under calm conditions the filaments, which contain gas vacuoles, accumulate at the surface, giving the appearance known to sailors as "sea sawdust". In littoral habitats, blue-green algae are usually abundant on rocks, on undisturbed sand and mud, in salt marshes, and mangrove swamps. Their tolerance of osmotic shock and of extremes of temperature, as well as their preference for reducing conditions, make them well suited for such habitats. In limestone areas endolithic forms such as *Entophysalis* and *Hyella* are abundant. Heterocystous marine forms are active in nitrogen fixation and, when abundant, add appreciably to the income of combined nitrogen of the habitat. *Calothrix crustacea,* an abundant algae of coral reef flats, is responsible for fixation at a rate of 1.8 kg N/h/day, which is the highest ever reported from a natural habitat, and evidently makes an important contribution to the productivity of the coral reef community.[319]

2. Nitrogen Fixation

The most intensive investigated aspects of soil algology has been asymbiotic nitrogen fixation.[314,315,320-338] Particular attention has been paid to the blue-green algae of rice soils. Information is available from several countries including India,[316,339-345] Japan,[346-349] Philippines,[350,351] Egypt,[352] Italy,[353] Australia,[354] Soviet Union,[336] Senegal,[355] Ivory Coast,[356] Nigeria,[357] Uganda,[358] Korea,[359] and Malaysia.[360]

Summarizing 30 years of studies on algal distribution in Japan, Southeast Asia, India,

Africa, Watanabe, and Yamamoto[361] reported that nitrogen-fixing species were widely distributed in both tropical and temperate areas, but were more abundant in the tropics. They comprise about 12% of the total algal flora in the tropics and about 2% in temperate zones north of 30° latitude. The most common tropical species belong to the genera *Anabaena, Aulosira, Nostoc, Scytonema,* and *Tolypothrix.*[342] Venkataraman[316] reported that only 33% of 2213 Indian soil samples contained dinitrogen fixers and while blue-green algae were widely distributed, certain geographical areas were rich and others relatively poor. For example, 87% of soils from Uttar Pradesh and 60% of soils from West Bengal contained nitrogen-fixing species, as compared with only 9 and 7% for Kerala and Kashmir. *Nostoc* and *Anabaena* were reported as universal, with species of *Calothrix, Cylindrospermum, Mastigocladus, Westiella,* and *Aulosira fertilissima* sporadically appearing as dominants. Singh[342] reported *Aulosira fertilissima* to be the most widespread and important blue-green algae associated with rice soils.

Nostoc commune, Nostoc sphaericum Vaucher, *Fischerella muscicola* (Thurent) Gomont, *Tolypothrix tenuis* Kutzing, and *Scytonema* spp. are typical inhabitants of undisturbed soils in the Soviet Union.[362,363] Moist soils contain *Nostoc paludosum* Kutzing, *Anabaena variabilis,* and *Calothrix* sp. Special strains of *Anabaena* characterize saline soils while *Nostoc muscorum* and *Nostoc punctiforme* are ubiquitous. Granhall[364] found 66% of 64 Swedish soils contain nitrogen-fixing blue-green algae and 47% contained nitrogen-fixing strains distributed among seven genera in temperate soils. Shtina,[325] who investigated 15 soils (including two tundra soils) in the Soviet Union, found dinitrogen-fixing algae to be universal, with from one (a virgin, montane soil), to 18 (cultivated, former woodland soil), or 19 (virgin sod-podzolic soil) species per soil. In steppe and desert soils nitrogen-fixing algae are important components of rain crusts, algal crusts, and lichen crusts.[363,365-374] Free-living nitrogen fixing blue-green algae and lichen phycobionts and moss-algal associations in cold dominated ecosystems are well documented.[363,375-377] In both tundra[378,379] and Antarctic[380,381] soils *Nostoc commune* is most common nitrogen fixers both as asymbiont and as the phycobiont in lichens. Due to variation in counting methods, direct comparisons of numbers presented by different scientists does not necessarily represent a true comparison of the relative importance of the algae. Estimates of the abundance of nitrogen-fixing algae range from zero (in some polar, temperate, and tropical soils) to 18 million filaments of *Cylindrospermum licheniforme* per gram of soil in India.[342] Shtina[325] reported 3.5 to 4.1 g of *Nostoc commune* per square meter on a rendzina soil, 200,000 *Nostoc commune* cells per square centimeter on chestnut, and 575,000 on dark chestnut soils. Up to 150 *Nostoc sphaeroides* colonies per square meter on alluvial soil and up to 310,000 *Cylindrospermum licheniforme* cells per square centimeter on sod-podzolic soils were also reported. While solonetz soils support up to 12 million cells per square centimeter, numbers do not usually exceed 1000 to 5000 cell per gram of soil in temperate soils in the Soviet Union.[382,383]

All of the samples from the nursery soil from Alaskan tundra on two different dates, contained blue-green algae.[384] Numbers ranged from 48,000 to 96,000/g on the surface, and from 1600 to 13,300/g for the 5 to 10 mm level. Nitrogen-fixing algae were not found in any of the nine Appalachian mountain soil samples, or in two out of five soils from the Piedmont, or in five of seven from the Atlantic Coastal Plain. Woodland soils with blue-green algae included two sites dominated by loblolly pine (with 130 and 640 algae/g), one site supporting a short leaf pine-red oak-hickory stand (1200 algae/g), one stream bottom site supporting beech (160 algae/g), and one soil dominated by cypress (16,500 algae/g). Sites in Alaska were reported to support from zero nitrogen-fixing algae per gram in birchheather and level grass dominated communities, to 6200/g on a soil of willow-moss community. Although the Alaskan soil with nitrogen-fixers ranged in pH from 5.8 to 6.5, it was reported that the soils from North Carolina (including the nursery) had a pH of less than 6.0. The sample from which the 96,000/g count originated had a pH of 5.8 and 13.2%

moisture by weight.[384] The most probable number of blue-green algal particles per gram of soil in eastern Washington fluctuated from 0.180/g, and from 67.0 to 160,000/g in cultivated and uncultivated silt loams respectively.[385,386]

Shields et al.[366] evaluated several surface crusts of arid lands in New Mexico for their Kjeldahl (total) N, No_2-N + NO_3-N, and amino-N in order to measure the nitrogen status of soil contents on three different dates. Amino nitrogen averaged 256 mg/kg of alga-stabilized gypsum crust compared to 10 mg/kg of soil 6 in. below the surface in June. Total NO_3 + NO_2-N in June averaged 5.6 mg/kg of crust, and 0.45 mg/kg of soil at 6 in. Differences among dates were insignificant. Algal crusts on leva soils averaged 1031 mg total N/kg, a value less than for lichens stabilized crusts (1553 mg/kg), but far greater than bare soil (225 mg/kg of soil). Algae are responsible for a difference in total nitrogen of from 77 to 133% in rice soils as compared with soils without algae.[387,388] Under fescue grass in the southeastern U.S., blue-green algal crusts include 2.76% total C and 0.18% total N compared with 0.67 and 0.06% on fallow soil without a crust.[389]

Nitrogen fixation rates in temperate and tropical soils are alike, with tropical rates ranging up to 69 kg nitrogen fixed per hectare during a 6 week period in pot culture with rice.[339] Yoshida et al.[351] estimated that up to 870 kg were fixed per hectare per year in Philippine soils, in pot culture experiments. Temperate soils with blue-green algae may fix up to 51 or 94 kg of nitrogen per hectare per year.[318,364,390] However, diurnal and seasonal fluctuations in activity of nitrogen-fixing blue-green algal populations are not well documented, and most estimates are based on quantifications made during the growing season. Maximal rates for temperate soils with blue-green algae during summer months, include 40 µg N_2 fixed/m^2/h in arable Scottish soils, 140 µg N_2/m^2/h in Scottish pasture,[314] and 0.06 to 25.7 mg/m^2/h[364] to 11 mg/m^2/day[390] in tilled and virgin Swedish soils and 6 to 10 g N/ha/h in soils of England and Wales.[391] Due in part to algal fixation and in part to bacterial fixation in dicot rhizospheres, the soils have accumulated about 2000 kg N per hectare, an average annual increment of 39 kg for the stubbed plot and 49 kg for the wooded plot.[392]

Estimates based on in vitro acetylene reduction assays varied greatly, ranging from 0.002 mg N/m^2/year for tundra soils of Devon Island, Canada, dominated by *Nostoc commune*,[379] to 11,500 mg N/m^2/year for a mire in Sweden dominated by various vascular species.[390] Estimates for soils of diverse habitats (heath, meadow, birchwood stands, snow) in Norway ranges from 100 to 250 mg N/m^2 year.[293] Morne[394] reported rates up to 2.4 mg N/m^2/year from Antarctica.

Henriksson et al.[395] reported that the temperate soils had a greater capacity (2.1 g N fixed/g soil/h) than the tropical soils (0.45 g N fixed/g soil/h). Maximum fixation rates for the soils were recorded at 20°C for Swedish samples and at 25°C for the Indian samples. Differences were not related to pH or nutrient, although the samples with the least activity had the greatest ionic strength as estimated by electrical-conductivity. It has, however, not been fully established that temperate soils support more active populations of nitrogen-fixing algae. Henriksson et al.[395] found values for virgin soils to be similar to wheat fields. Others[396,397] contend that dinitrogen fixation by asymbionts is not significant in either tilled or virgin grassland soils in North America.[396,397] Woodmansee[398] found that the blue-green algae might be important in steppe zones in either cultivated or uncultivated soils. Numbers of heterocystous algae were reduced by farming practices in eastern Washington,[386] while in contrast, temperate soils in the Soviet Union tended to increase in numbers of nitrogen-fixation algae upon cultivation.[399]

Limited fixation by blue-green algae in cultivated soils and laboratory culture was due to the introduction of combined sources of nitrogen via fertilization and irrigation.[363,383,389,391,400-407] For example, aliphatic amines up to 6 C long inhibited O_2 production and nitrogen fixation by *Anabaena subcylindrica* as much as 2500 times than did NH_3.[408] However, evidence for the hypothesis that this might be a nonspecific salt effect has come from soil[395] and laboratory data.[330,409] On the other hand, Shtina[362] showed that an increase in growth of potential nitrogen-fixing algae accompanied application of nitrogen alone, or in combination with K or P.

Nitrogen fixation in blue-green algae is dependent on light as a source of energy.[410-413] Although *Nostoc muscorum* and *Anabaenopsis circularis*[349] and other species[414,415] can fix nitrogen in the dark with glucose as an energy source, nitrogenase activity has never been detected from below the surface of a soil on which algae are active fixers.[395] *In situ* acetylene reduction assays have demonstrated that nocturnal fixation in rice soils and subtropical stands of *Pennisetum* follows active photosynthesis during the day.[416] Fluctuations in nitrogen fixation rates by soil algae in Morocco coincided with diurnal changes which might have been inhibited midday by irradiance of 50,000 lx[417] while some strains isolated from northern Sweden did not have nitrogenase activities saturated by light.[364] Nitrogen fixation by mats of blue-green algae in *Pennisetum* was also inhibited by high irradiance. An Egyptian strain of *Nostoc commune*, grown in laboratory culture, showed optimum growth and fixation at 6000 lx.[418,419] Fixation by algal crusts in the American Great Basin Steppe was optimal at 200 microEs/cm^2/sec.[420]

Swedish nitrogen fixation was also influenced by temperature.[39] Samples collected from Scottish soils in the winter, fixed nitrogen in the laboratory at 25°C but not at 4°C.[314] Whereas *Westiellopsis prolifica* showed greatest growth at 40°C in laboratory studies, nitrogen fixation rates were maximal between 30 and 35°C.[421] Both predawn low temperatures and midday highs suppressed *in situ* fixation by blue-greens associated with *Pennisetum*.[422] Adaptation to low temperatures by algae of high latitudes was responsible for fixation of significant amounts of nitrogen.[375,394,402,423]

In both tropical[316,342] and temperate zones[383,424] greater growth of nitrogen fixing blue-green algae occurred in moist soils. Dried Broadbalk soils, when rewetted in the laboratory, showed nitrogenase activity within 2 hr.[392] Activity of *Nostoc muscorum* crusts following desiccation was, proportional to the degree to which they were rewetted.[425] Growth of *Scytonema ocellatum* in lawns and grasslands in India was greatest at 40% of the soil's water holding capacity[426] and while optimal fixation in *Pennisetum* swards was between 22 and 42%[422] most species showed optimum nitrogen fixation between 80 and 100%.[342,383,420,427]

The distribution and nitrogen-fixation activities of blue-green algae was limited considerably by pH changes in soil.[314,363,399,428-430] Blue-green algae were abundant on acidic soils than in neutral to alkaline soils.[424,431] In general, optimal level of pH for growth and nitrogen fixation were neutral to alkaline.[314] Optimal pH ranges for some algae were 6.6 to 10.0 for *Aphanothece* sp.,[432] 7.4 to 9.0 for species of *Anabaena*.[410,433,434] However, the pH alone was not only the limiting factor for nitrogenase activity.[435] The pH optimum for *Nostoc punctiforme* was found to be 7.6 but nitrogenase activity was detected in culture from pH 5 to 10.5.[410] Other factors which influenced blue-green algae nitrogenase activity were ionic strength and oxidation-reduction potential. High levels of O_2 inhibit fixation, especially by nonheterocystous blue-green algae.[333] Reducing conditions in waterlogged rice soils increase nitrogen fixation.[342]

Increased phosphorus in soil and freshwater systems stimulated fixation which might be attributed to increased ATP synthesis[332,436] which was required for nitrogenase activity. Stimulation has been noted for algae in pot cultures with rice,[339] in field experiments with $CaCO_3$ + K_2HPO_4[340,346,406] in lake waters with inorganic and detergent phosphates[332] and in soils with insoluble $CaPO_4$.[340] Algal nitrogenase activity in pure culture was also dependent upon the supply of Mo, Ca, and Na, and that Co is required for growth.[395,346,437-439] Molybdenum was not required for NO_3^- assimilation by *Anabaena cylindrica* or *Cylindrospermum sphaericum*,[440,441] although its presence promoted healthier growth. Maximal growth of *Anabaena* with NO_3^- as the nitrogen source occurred at 0.075 to 0.100 ppm Mo, a value equal to about one third of the optimal requirement with N_2 as the source.[440] Jacobs and

Lind[442] found the optimal concentration of Mo to vary with temperature, being greatest (50 $\mu g/\ell$) at 30°C and less at 15°C (15 $\mu g/\ell$) and 23°C (5 $\mu g/\ell$). Molybdenum concentration did not affect growth with NH_4Cl.[440] Sodium was required in amounts greater than 5 ppm for optimal fixation in the absence of combined nitrogen.[437,439] Vanadium could not replace Mo as it could to some extent in *Azotobacter* nor can Sr replace Ca.[438] Calcium content of soil may be important for improvement of the quality of the ionic environment by alleviating the deleterious effects of high concentrations of salts through precipitation of excess anions.

The addition of $CaCO_3$ stimulated the nitrogenase activity and algal growth in rice soils.[346] In laboratory culture, nitrogen fixation by *Aulosira fertilissima* was stimulated by addition of $CaCO_3$, KH_2PO_4, or $CaCO_3$ + KH_2PO_4 but was depressed when all three were simultaneously incorporated into the medium.[406] Liming of soil fescue grass increased nitrogen fixation in the southeastern U.S.[389] Calcium rich soils in the Soviet Union also tended to support more diverse and active populations of nitrogen-fixing blue-green algae.[383]

Algal populations in cultivated soils have been manipulated for three different, yet interrelated purposes. These include inoculation of paddy soils with nitrogen-fixing species for the purpose of supplying combined nitrogen, the use of aquatic and edaphic algal growth as green manure, and the inoculation of unproductive soils with green or blue-green algae in order to effect reclamation of the soil through improved structure. The use of algae for the purpose of supplying combined nitrogen has become common during recent years. The results of inoculation of rice soils with nitrogen-fixing species of blue-green algae have been summarized.[316,443] Increased yields between 5 and 30% have been reported from China,[444-446] Vietnam,[447] the Philippines,[350] Japan,[448-450] Egypt,[451] Soviet Union,[452,453] and India. In India, Subramanyan and co-workers[454-460] tested the influence of different algal mixtures on yields of grain and straw in field tests over a 5 year period. No differences among varieties of rice were noted, when algae inoculum was applied at rates 6 equivalent to 200 g/ha(wet wt). Results were compared with uninoculated fields, fields fertilized with lime plus superphosphate plus Na-molybdate, fields fertilized with $(NH_4)_2SO_4$, and fields fertilized with the first three chemicals and inoculated with algae. Also assessed were the effects of rabbing (burning) as a measure of control of wild populations of algae. They found that rabbing altered the floristic make-up and dominance patterns within the algal communities without affecting rice yields. *Anabaena* dominated all plots except the uninoculated control, which was dominated by *Cylindrospermum* sp., and the burned, and fertilized plots were dominated by species of *Rivularia* and *Gloeotrichia*. Similar results were obtained by Jagnow[461] from experiments in which K_2HPO_4 and Na_2MoO_4 were employed.

The grain and straw yields in plots fertilized with *Nostoc*, *Tolypothrix* sp., and *Westiella* sp. plus lime plus superphosphate plus Na-molybdate, were greater than for plots fertilized with urea plus lime plus superphosphate plus Na-molybdate.[462] Similar reports of improved yields of rice in India due to algalization include those of Jha et al.[463] in which a 36% increase was attributed to *Tolypothrix tenuis*, and of Singh[342] who reported a 114.8% increase in yields upon inoculation with *Aulosira fertilissima*.

In Japan, large scale trials at 11 experimental stations over a 5 year period showed that inoculation with *Tolypothrix tenuis* resulted in positive annual increments in the yield of rice of 2.7, 8.4, 19.1, and 21.8% on the average.[449,464,465] However, problems were associated with mass culture, distribution, and inoculation.[316,345,405,448,464,465]

The enzyme system of blue-green algae (nitrogenase) by which nitrogen fixation is effected, is generally similar to that occurring in nitrogen fixing bacteria, and one of the principal points of interest is the manner in which this oxygen-sensitive enzyme is protected from oxygen. It is now clear that in the most active and abundant nitrogen fixing species, such as *Anabaena* and *Nostoc*, this is accomplished by nitrogen fixation being localized in the heterocysts,[466] and it appears that all algae possess heterocysts which fix nitrogen. Many kinds of evidences show that under aerobic conditions nitrogen fixation in these species is carried out in these specialized cells. Ammonium nitrogen, the first stable product of nitrogen fixation, suppresses the formation of heterocysts.[467] If the algal material is then transferred to ammonium free medium, heterocysts are differentiated, and at the same time nitrogenase activity develops. Heterocysts isolated from vegetative cells by lysozyme treatment will reduce acetylene (an index of nitrogen fixation) if inoculated with dithionite as a source of hydrogen and ATP as a source of energy and autoradiographic experiments with the short-lived radioactive isotope ^{13}N indicated heterocysts as the loci of nitrogen fixation in intact filaments.[468] Heterocysts lack phycocyanin and other components of photosystem II, and so are not capable of complete photosynthesis and oxygen production. Cytochemical tests in fact show mature heterocysts to be strongly reducing, compared to vegetative cells, and they are thus suited to be the site of nitrogenase activity. Possibly the thick heterocyst envelop with its high contents of lipid plays a part in restricting the inward diffusion of oxygen from the environment. The heterocysts retain photosystem I and the ability to produce ATP by photosynthetic phosphorylation. ATP from this source is evidently the main supply for nitrogen fixation, since the rate depends on light intensity in much the same way as does photosynthesis, although blue-green algae can fix nitrogen slowly in the dark. The other prerequisites for nitrogen fixation are a hydrogen donor and carbon skeleton for converting the ammonia produced to organic form. Since heterocysts cannot assimilate carbon dioxide photosynthetically, the ultimate source of both hydrogen and carbon dioxide must be the vegetative cells. Radioautography of filaments supplied with ^{14}C-carbon dioxide in light shows that carbon fixed in the vegetative cells is, indeed translocated into heterocysts.[317] By the same token, the vegetative cells must be dependent on heterocysts for fixed nitrogen. This has been demonstrated indirectly by making use of the fact that the pigment phycocyanin acts as a nitrogenous reserve. If a filament that has been starved of nitrogen, and hence is low in phycocyanin, is allowed to fix nitrogen, the phycocyanin can be observed by microspectrophotometry to reappear, first in the cells adjacent to the heterocysts, and later spreading outwards to cells remote from heterocysts. If the starved filament is provided with a fixed nitrogen source, such as nitrate, the phycocyanin reappears in all cells at the same time.[469]

Some filamentous nonheterocystous algae, such as *Plectonema boryanum*, *Phormidium*, and *Oscillatoria* species, are also able to fix nitrogen under anaerobic and microaerophilic conditions.[470,471] The capacity rapidly disappears if the algae is exposed to free oxygen. It seems probable that, likewise, all cells of *Anabaena* or *Nostoc*, and not only the heterocysts are able to fix nitrogen under anaerobic conditions. A problem is presented by the marine planktonic from *Trichodesmium*, occurs in a oxygenated environment. However, the filaments of *Trichodesmium* occur in bundles, and it may be that nitrogen fixation carried out in the filaments near the center of the bundle, where the oxygen supply is restricted. A unicellular form *Gloecapsa* sp., has also been shown to fix the nitrogen[472] but it appears to do this only during a short period of the growth cycle. This ability to fix the nitrogen is more widespread among blue-green algae than was suspected a few years ago, and that species dismissed as nonnitrogen fixers must be re-examined under anaerobic conditions. Nevertheless, there do appear to be species, e.g., the common plankton form *Microcystis aeruginosa*, that are not able to fix nitrogen under any condition.

3. Blue-Green Algae Symbiosis

Blue-green algae have remarkable capacity for forming associations with other organisms. Although not as frequently associated with ascomycetous fungi in lichens as green algae, species of *Calothrix*, *Hyella*, *Nostoc*, *Scytonema*, and of other genera of blue-green algae, nevertheless occur in this form of symbiosis. A different type of symbiosis with a fungus occurs in *Geosiphon* in which *Nostoc* occurs intracellulary in a phycomycete. Another blue-green algae, *Richelia intracellularis* occurs in the cells of the marine plankton diatom

Rhizosolenia. Pockets of *Nostoc* and *Anabaena* occur in the tissue of bryophytes, such as *Anthoceros* and *Blasia* of the aquatic fern *Azolla,* of the cycads *Macrozamia,* and other genera, and of the agiosperm *Gunnera. Aphanocapsa* occurs in various sponges, both extracellularly and intracellularly. Association with protozoa-such as *Cynophora paradoxa, Glucocystics nostochinearum, Paulinella chromatophora* are of particular interest, since blue-green partners are intracellular, devoid of cell wall, and difficult to assign to any free living species. Certain of these ''Cyanelles'' have been isolated and cultured. In many of these associations, e.g., the lichens and the protozoa with cyanelles the nonalgal partner certainly benefit from photosynthetic products. However, in the lichens, the benefit from nitrogen fixed by algae is also important, and in the cycads and *Gunnera* it is predominant, since the algal partner in these appears to grow at least partially at the expense of organic substrates provided by the higher plants.

The appearance of *Anabaena-Azolla* associations in their natural habitat are strongly correlated with surface and bottom temperatures of the pond.[473,474] The optimum temperature for growth is 20°C or above. High light intensity is injurious to growth of the *Azolla.* Fern grows well under reduced light conditions and at pH 5 to 7. Laboratory studies on the factors affecting *Anabaena-Azolla* growth also revealed that a temperature of 23°C day to 18°C night was optimal for growth.

The capability of the *Anabaena-Azolla* symbiosis to fix atmospheric nitrogen by virtue of the nitrogenase enzyme in the phycobiont is now well documented.[473-477] The symbiosis of *Azolla* and *Anabaena* is the only known association of a pteridophyte with a cynophore. *Anabaena azolla* is found in all species of the aquatic heterosporous fern *Azolla.* It is enclosed within a small, basal cavity of the upper green lobe of the bilobed, alternate leaves. Very rapid propagation of the in fact *Azolla* is generally achieved by fragmentation. In many-temperature and tropical regions, *Azolla* is considered a pest because it can cover a farm pond or rice paddy rapidly.[478] However, it has long been cultivated as a green manure for rice crops in Southeast Asia. This latter use provides the basis for the recent intense interest in this unique symbiosis. The algal symbiont is a nitrogen-fixing organism that behaves much like *Rhizobium* symbiont and legumes. In addition, it is also established that principal enzyme governing utilization of nitrate is nitrate reductase which also functions in the host member of symbiosis.[473] The total nitrogen content assimilated by symbiosis when grown in air and on nitrate containing medium would be the sum of the functioning of the two enzymatic systems. The system is similar to legume-*Rhizobium* symbiosis. Prominent enzymes involved in nitrogen fixation in this system are nitrate reductase and nitrogenase.

Various environmental factors influence the activities of nitrate reductase and nitrogenase of *Anabaena-Azolla* symbiosis. Nitrate reductase activity is generally unaffected by ammonium nitrogen, oxygen or varying pH.[474] However, nitrogenase is more sensitive to osmotic stress, ammonium-nitrogen, and oxygen.

REFERENCES

1. **Eylar, O. R. and Schmidt, E. L.,** A survey of heterotrophic microorganisms from soil for ability to form nitrite and nitrate, *J. Gen. Microbiol.,* 20, 473—481, 1959.
2. **Hirsch, P., Overrein, L., and Alexander, M.,** Formation of nitrite and nitrate by actinomycetes and fungi, *J. Bacteriol.,* 82, 442—448, 1961.
3. **Verstraete, W. and Alexander, M.,** Heterotrophic nitrification by *Arthobacter* sp., *J. Bacteriol.,* 110, 955—961, 1972.
4. **Verstraete, W. and Alexander, M.,** Mechanism of nitrification by *Arthobacter* sp., *J. Bacteriol.,* 110, 962—967, 1972.
5. **Verstraete, W. and Alexander, M.,** Heterotrophic nitrification in samples of natural ecosystems, *Environ. Sci. Technol.,* 7, 39—42, 1973.

6. **Giddens, J.,** Nitrogen Cycling in Georgia Soils, Research Bulletin, University of Georgia, Athens, Ga., No. 327, 1—37, 1985.
7. **Focht, D. D. and Verstraete, W.,** Biochemical ecology of nitrification and denitrification, *Adv. Microb. Ecol.,* 135—214, 1977.
8. **Alexander, M., Marshall, K. C., and Hirsch, P.,** Autotrophy and heterotrophy in nitrification, Trans. 7th Int. Congr. Soil Sci., Madison, Wis., 2, 586—591, 1960.
9. **Alexander, M.,** Nitrification, in *Soil Nitrogen,* Bartholomew, M. V. and Clark, F. E., Eds., Am. Soc. of Agron., Madison, Wis. 1965, 307—343.
10. **Painter, H. A.,** A review of literature on inorganic nitrogen metabolism in microorganisms, *Water Res.,* 4, 393—450, 1970.
11. **Walker, N.,** Nitrification and nitrifying bacteria, in *Soil Microbiology,* Walker, N., Ed., New York, 1978, 133—146.
12. **Belser, L. W. and Schmidt, E. L.,** Nitrification in soils, in *Microbiology,* Schlessinger, D., Ed., Am. Soc. for Microbiol., Washington, D.C., 1978, 348—351.
13. **Belser, L. W. and Schmidt, E. L.,** Diversity in the ammonia-oxidizing nitrifier population of a soil, *Appl. Environ. Microbiol.,* 36, 584—588, 1978.
14. **Bock, E.,** Lithoautotrophic and chemoorganotrophic growth of nitrifying bacteria, in *Microbiology,* Schlessinger, D., Ed., Am. Soc. for Microbiol., Washington, D.C., 1978, 310—314.
15. **Hooper, A. B.,** Nitrogen oxidation and electron transport in ammonia oxidizing bacteria, in *Microbiology,* Schlessinger, D., Ed., Am. Soc. for Microbiol., Washington, D.C., 1978, 299—304.
16. **Knowles, R.,** Common intermediates of nitrification and denitrification and the metabolism of nitrous oxide, in *Microbiology,* Schlessinger, D., Ed., Am. Soc. for Microbiol., Washington, D.C., 1978, 367—371.
17. **Nicholas, D. J. D.,** Intermediary metabolism of nitrifying bacteria, with particular reference to nitrogen, carbon and sulfur compounds, in *Microbiology,* Schlessinger, D., Ed., Am. Soc. for Microbiol., Washington, D.C., 1978, 305—309.
18. **Schmidt, E. L.,** Nitrifying microorganisms and their methodology, in *Microbiology,* Schlessinger, D., Ed., Am. Soc. for Microbiol., Washington, D.C., 1978, 228—291.
19. **Verstraete, W.,** Heterotrophic nitrification in soils and aqueous media, *Izv. Akad. Nauk SSSR Ser. Biol.,* 4, 541—558, 1975.
20. **Yamafugi, K., Kondo, H., and Omura, H.,** Distribution of oxime in plant and animal tissues, *Enzymologia,* 14, 153—156, 1950.
21. **Emery, T.,** Biosynthesis and mechanisms of action of hydroxamate-type siderochromes, in *Microbial Metabolism,* Neilands, J. B., Ed., Academic Press, New York, 1974, 107—123.
22. **Leong, S. A. and Neilands, J. B.,** Siderophore production by phytopathogenic microbial species, *Arch. Biochem. Biophys.,* 218, 351—359, 1982.
23. **Akers, H. A.,** Multiple hydroxamic acid microbial iron chelators (siderophores) in soils, *Soil Sci.,* 135, 156—159, 1983.
24. **Powell, P. E., Cline, G. R., Ried, C. P., and Szaniszlo, P. J.,** Occurrence of hydroxamate siderophore iron chelators in soils, *Nature (London),* 287, 833—834, 1980.
25. **Castignetti, D. and Gunner, H. B.,** Sequential nitrification by an *Alcaligenes* sp. and *Nitrobacter agilis.,* *Can. J. Microbiol.,* 26, 1114—1119, 1980.
26. **Castignetti, D. and Gunner, H. B.,** Nitrite and nitrate synthesis from pyruvic oxime by an *Alcaligenes* sp., *Curr. Microbiol.,* 5, 379—384, 1980.
27. **Castignetti, D. and Hollocher, T. C.,** Vigorous denitrification by a heterotrophic nitrifier of the genus *Alcaligenes, Curr. Microbiol.,* 6, 247—252, 1981.
28. **Castignetti, D. and Hollocher, T. C.,** Nitrogen redox metabolism of a heterotrophic nitrifying-denitrifying *Alcaligenes* sp. from soil, *Appl. Environ. Microbiol.,* 44, 923—928, 1982.
29. **Castignetti, D., Petithory, J. R., and Hollocher, T. C.,** Pathway of oxidation of pyruvic oxime by a heterotrophic nitrifier of the genus *Alcaligenes:* evidence against hydrolysis to pyruvate and hydroxylamine, *Arch. Biochem. Biophys.,* 224, 587—593, 1983.
30. **Amarger, N. and Alexander, M.,** Nitrite formation from hydroxylamine and oximes by *Pseudomonas aeruginosa, J. Bacteriol.,* 95, 1651—1657, 1968.
31. **Obaton, M., Amarger, N., and Alexander, M.,** Heterotrophic nitrification by *Pseudomonas aeruginosa,* *Arch. Mikrobiol.,* 63, 122—132, 1968.
32. **Castignetti, D. and Hollocher, T. C.,** Heterotrophic nitrification among denitrifiers, *Appl. Environ. Microbiol.,* 47, 620—623, 1984.
33. **Clark, K. G. and Gaddy, V. L.,** Composition and nitrification characteristics of some sewage and industrial sludges, *Farm Chem.,* 118, 41—45, 1955.
34. **Premi, P. R. and Cornfield, A. H.,** Incubation study of nitrification of digested sewage sludge added to soil, *Soil Biol. Biochem.,* 1, 1—4, 1969.
35. **Premi, P. R. and Cornfield, A. H.,** Incubation study of nitrogen mineralization in soil treated with dried sewage sludge, *Environ. Pollut.,* 2, 1—4, 1971.

36. **King, L. D.**, Mineralization and gaseous loss of nitrogen in soil-applied liquid sewage sludge, *J. Environ. Qual.*, 2, 356—358, 1973.
37. **Ryan, J. A., Keeney, D. R., and Walsh, L. M.**, Nitrogen transformations and availability of an anaerobically digested sewage in soil, *J. Environ. Qual.*, 2, 489—492, 1973.
38. **Wilson, D. O.**, Nitrification in soil treated with domestic and industrial sewage sludge, *Environ. Pollut.*, 12, 73—82, 1977.
39. **Timmer-Ten Hoor, A.**, A new type of thiosulphate oxidizing, nitrate reducing microorganisms: *Thiomicrospira denitrificans* sp., *Neth. J. Sea Res.*, 9, 343—353, 1975.
40. **Timmer-Ten Hoor, A.**, Energetic aspects of the metabolism of reduced sulphur compounds in *Thiobacillus denitrificans*, *Antonie van Leeuwenhoek*, 42, 483—492, 1976.
41. **Timmer-Ten Hoor, A.**, Denitrificerende Kleurloze Zwavelbacterien, Ph.D. thesis, State University of Netherland, Groningen, 1977.
42. **Pfitzner, J. and Schlegel, H. G.**, Denitrification bei *Hydrogenomonas eutropha* Stam, H16, *Arch. Mikrobiol.*, 90, 199—211, 1973.
43. **Bovell, C.**, The effect of sodium nitrite on the growth of *Micrococcus denitrificans*, *Arch. Mikrobiol.*, 59, 13—19, 1967.
44. **Verhoeven, W., Koster, A. L., and van Nievelt, M. C. A.**, Studies on true dissimilatory nitrate reduction. III. *Micrococcus denitrificans* Beijerincki, a bacterium capable of using molecular hydrogen in denitrification, *Antonie van Leewenhoek*, 20, 273—284, 1954.
45. **Ishaque, M., Donawa, A., and Aleem, M. I. H.**, Energy-coupling mechanisms under aerobic and anaerobic conditions in autotrophically grown *Pseudomonas saccharophilia*, *Arch. Biophys.*, 159, 570—579, 1973.
46. **Auling, G., Reh, M., Lee, C. M., and Schlegel, H. G.**, *Pseudomonas pseudoflava*, a new species of hydrogen-oxidizing bacteria; its *differentiation* from *Pseudomonas flava* and other yellow pigmented Gram-negative, hydrogen-oxidizing species, *Int. J. Syst. Bacteriol.*, 28, 82—95, 1978.
47. **Attwood, M. M. and Harder, W.**, A rapid and specific enrichment procedure for *Hyphomicrobium* spp., *Antonie van Leeuwenhoek*, 38, 369—378, 1972.
48. **Sperl, G. J. and Hoare, D. S.**, Dentrification with methanol: a selective enrichment for *Hyphomicrobium* species, *J. Bacteriol.*, 108, 733—736, 1971.
49. **Buchanan, R. E. and Gibbons, N. E.**, *Bergey's Manual of Determinative Bacteriology*, 8th ed., Williams & Wilkins Baltimore, 1974.
50. **Van Gent-Ruikters, M. L. W., de Vries, W., and Stouthamer, A. H.**, Influence of nitrate on fermentation pattern, molar growth yields and synthesis of cytochrome in *Propionibacterium pentosaccum*, *Gen. Microbiol.*, 88, 36—48, 1975.
51. **Satoh, T., Hoshino, Y., and Kitamura, H.**, Isolation of denitrifying photosynthetic bacteria, *Agric. Biol. Chem.*, 38, 1749—1751, 1974.
52. **Satoh, T., Hoshino, Y., and Kitmura, H.**, *Rhodopseudomonas sphaeroides*, forma sp. denitrificans, a denitrifying strain as a subspecies of *Rhodopseudomonas sphaeroides*, *Arch. Microbiol.*, 108, 265—269, 1976.
53. **Pyne, W. J., Rowe, J. J., and Sherr, B. F.**, Dentrification a plea for attention, in *Nitrogen Fixation*, Newton, W. E. and Orme-Johnson, W. H., Eds., University Park Press, Baltimore, 1980, 29—42.
54. **Starkey, R. L.**, Interrelations between microorganisms and plant roots in the rhizosphere, *Bacteriol. Rev.*, 22, 154—174, 1958.
55. **Woldendrop, J. W.**, The influence of living plants on denitrification, *Meded. Landbhogesch. Wageningen*, 63, 1—100, 1963.
56. **Garcia, J. L.**, Influence de la rhizosphère du riz sur Pactivite denitrifiante potentlalle des sols de rizieres du Senegal, *Oecol. Plant*, 8, 315—323, 1973.
57. **Garcia, J. L., Raimbault, M., Jacq, V., Rinaudo, G., and Roger, P.**, Activites microbines dans les sols de rizieres du Senegal relations avec les caracteristiques physicochimiques et influence de la rhizosphère, *Rev. Ecol. Biol. Soil*, 11, 169—185, 1974.
58. **Raimbault, M.**, Etude de l'influence inhibitrice de l'acetylene sur la formation biologique du methane dans un sol de riziere, *Ann. Microbiol.*, 126A, 247—258, 1975.
59. **Stefanson, R. C.**, Rekative rates of denitrification and nonsymbiotic nitrogen fixation in the soil-plant systems, *Soil Biol. Biochem.*, 5, 869—880, 1973.
60. **Hall, J. B.**, Nitrate-reducing bacteria, in *Microbiology*, Schlessinger, D., Ed., Am. Soc. for Microbiol., Washington, D.C., 1978, 296—298.
61. **Garcia, J. L., Pichinoty, F., Mandel, M., and Greenway, B.**, A new denitrifying saprophyte related to *Pseudomonas picketti*, *Ann. Microbiol.*, 128, 229—237, 1977.
62. **Pichinoty, F., Garcia, J. L., Job, C., and Durand, M.**, La denitrification chez *Bacillus licheniformis*, *Can. J. Microbiol.*, 24, 45—49, 1978.
63. **Okerele, G. U.**, Utilization and Production of N_2O by Denitrifiers Isolated from Different Soil Environments and the Effect of pH on the Rates and Products of Denitrification, M.S. thesis, Michigan State University, East Lansing, Mich., 1978.

64. **Gamble, R. N., Betlach, M. R., and Tiedje, J. M.,** Numerically dominant denitrifying bacteria from world soils, *Appl. Environ. Microbiol.,* 33, 926—939, 1977.
65. **Dowdell, R. J., Smith, K. A., Crees, R., and Restall, S. W. F.,** Field studies of ethylene in the soil atmosphere-equipment and preliminary results, *Soil Biol. Biochem.,* 4, 325—331, 1972.
66. **Paul, E. A., Myres, R. J. K., and Rice, W. A.,** Nitrogen fixation in grassland and associated cultivated ecosystems, *Plant Soil,* Special volume, 495—507, 1971.
67. **Pilot, L. and Patrick, W. H.,** Nitrate reduction in soils: effects of soil moisture tension, *Soil Sci.,* 114, 312—316, 1972.
68. **Stefanson, R. C.,** Soil denitrification in sealed soil plant systems. I. Effect of plants, soil, water content and soil organic matter content, *Plant Soil,* 37, 113—127, 1972.
69. **Van Schreven, D. A.,** Nitrogen transformations in the former subaqeuous soils of polders recently reclaimed from lake Ijssel. II. Losses of nitrogen due to denitrification and leaching, *Plant Soil.,* 18, 163—175, 1963.
70. **Russel, E. W.,** *Soil Conditions and Plant Growth,* 9th ed., Longmans, London, 1961, 311-317.
71. **Myres, R. J. K. and McCarity, J. W.,** Denitrification in undisturbed cores from a solodized solonetz horizon, *Plant Soil,* 37, 81—89, 1972.
72. **McCarity, J. W. and Myres, R. J. K.,** Denitrifying activity in solodized, Solonetz soils in eastern Australia, *Soil Sci. Soc. Am., Proc.,* 32, 812—817, 1968.
73. **Greenwood, D. J.,** Nitrification and nitrate dissimilation in soil. II. Effect of oxygen concentration, *Plant Soil,* 17, 378—391, 1962.
74. **Lance, J. C.,** Nitrogen removal by soil mechanisms, *J. Water Pollut. Control Fed.,* 44, 1352—1361, 1972.
75. **Chen, R. L., Kenney, D. R., Konrad, J. G., Holdings, A. J., and Graetz, D. A.,** Gas production in sediments of Lake Mendota, Wisconsin, *J. Environ. Qual.,* 1, 155—157, 1972.
76. **Dawson, R. N. and Murphy, K. L.,** The temperature dependency of biological denitrification, *Water Res.,* 6, 71—83, 1972.
77. **Goering, J. J. and Cline, J. D.,** A note on denitrification in seawater, *Limnol. Oceanogr.,* 15, 306—309, 1970.
78. **McCarty, P. L., Beck, L., and Amant, P. St.,** Biological denitrification of agricultural wastewaters by addition of organic materials, *Purdue Univ. Eng. Ext. Ser.,* 135, 1271—1285, 1969.
79. **Payne, W. J. and Riley, P. S.,** Suppression by nitrate of enzymatic reduction of nitric oxide, *Proc. Soc. Exp. Biol. Med.,* 132, 258—260, 1969.
80. **Parkasam, T. B. S. and Loehr, R. C.,** Microbial nitrification and denitrification in concentrated wastes, *Water Res.,* 6, 859—869, 1972.
81. **Sperl, G. T. and Hoare, D. S.,** Denitrification with methanol selective enrichment for *Hyphomicrobium* species, *J. Bacteriol.,* 108, 733—736, 1971.
82. **Taylor, B. F. and Heeb, M. J.,** The anaerobic degradation of aromatic compounds by a denitrifying bacterium, *Arch. Mikrobiol.,* 83, 165—171, 1972.
83. **Taylor, B. F., Compbel, W. L. and Chinoy, I.,** Anaerobic degradation of the benzene nucleus by facultative anaerobic microorganisms, *J. Bacteriol.,* 102, 430—437, 1970.
84. **Walsh, L. and Kenney, D. R.,** The pollution problem. I. Understanding the nitrogen cycle, *Hoards Dairyman,* 115, 732, 1970.
85. **Stewart, B. A.,** A look at agriculture practices in relation to nitrate accumulation, *Soil Sci. Spec. Publ. Ser.,* 4, 47—60, 1970.
86. **Trolldenier, G.,** Einfluss der Strickstoff and Kaliumonahrung von Weizen sowei dei Sauerstoffversorgung der Wurzeln auf Bakterienzahl, Wurzelatmung and Denitrifikation in der Rhizosphare, *Zentralbe. Bakteriol. Parasitenkd. Infektionskr. Abt.,* 126, 130—141, 1971.
87. **Ilyaletdinov, A. N. and Suleimenova, S. I.,** Effect of soil flooding terms on denitrification and mineralization of organic matter, *Mikrobiologia,* 40, 608—612, 1971.
88. **Downey, R. J.,** Nitrate reductase and respiratory adaption in *Bacillus stearothermophillus, J. Bacteriol.,* 91, 634-641, 1966.
89. **Itagaki, E., Fujita, T., and Sato, R.,** Cytochrome b_1-nitrate reductase interaction in a solubilized system from *Escherichia coli, Biochim. Biophys. Acta,* 51, 390—392, 1961.
90. **Kiszkiss, D. F. and Downey, R. J.,** Localization and solubilization of the respiratory nitrate reductase of *Bacillus stearothermophilus, J. Bacteriol.,* 109, 803—810, 1972.
91. **Najjar, V. A. and Chung, C. W.,** Enzymatic steps in denitrification, in *Symp. on Inorganic Nitrogen Metabolism,* McElroy, W. D. and Glass, B. H., Eds., Johns Hopkins Press, Inc., Baltimore, 1956, 260—296.
92. **Nason, A.,** Symposium on metabolism of inorganic compounds. II. Enzymatic pathways of nitrate, nitrite, and hydroxylamine metabolisms, *Bacteriol. Rev.,* 26, 16—41, 1962.
93. **Ruiz-Herrera, J. and DeMoss, J. A.,** Nitrate reductase complex of *Escherichia coli* K-12-participation of specific formate dehydrogenase and cytochrome b_1 components in nitrate reduction, *J. Bacteriol.,* 99, 720—729, 1969.

94. **Showe, M. K. and DeMoss, J. A.,** Localization and regulation of synthesis of nitrate reductase in *Escherichia coli, J. Bacteriol.,* 95, 1305—1313, 1968.
95. **Van'T Riet, J., Knook, D. L., and Planta, R. J.,** The role of cytochrome b₁ in nitrate assimilation and nitrate respiration in *Klebsiella aerogenes, FEBS Lett.,* 23, 44—46, 1972.
96. **Yamada, T. and Sakaguchi, K.,** Nitrogen fixation associated with a hot spring green alga, *Arch. Microbiol.,* 124, 161—167, 1980.
97. **Stewart, W. D. P. and Gallon, J. R., Eds.,** *Nitrogen Fixation,* Academic Press, New York, 1980.
98. **Gibson, A. H. and Newton, W. E., Eds.,** Current perspectives in nitrogen fixation,, Proc. 4th Int. Sym. Nitrogen Fixation, Australian Academy of Science Canberra, Canberra, 1981.
99. **Postgate, J. R.,** Prequisites for biological nitrogen fixation in heterotrophic bacteria, in *The Biology of Nitrogen Fixation,* Quispel, A., Ed., North-Holland, Amsterdam, 1974, 663—686.
100. **Postgate, J. R.,** Microbiology of the free-living nitrogen fixing bacteria excluding cyanobacteria, in *Current Perpectives in Nitrogen Fixation,* Gibson, A. H. and Newton, W. E., Eds., Australian Academy of Science, 1981, 217—228.
101. **Postgate, J. R.,** Biology of nitrogen fixation, fundamentals, *Philos. Trans. R. Soc. London, Ser.,* B, 296, 375—385, 1982.
102. **Mishustin, E. N. and Shilanikova, V. K.,** Free-living heterotrophic nitrogen-fixing bacteria, in *The Biology of Nitrogen Fixation,* Quispel, A., Ed., North-Holland, Amsterdam, 1974, 37—85.
103. **Line, M. A. and Loutit, M. W.,** Occurrence of *Azotobacter* in some soils of South Island, New Zealand, *N.Z. J. Agric. Res.,* 12, 630—638, 1969.
104. **Abd-el-Malek, Y.,** Free living nitrogen-fixing bacteria in Egyptian soils and their possible contribution to soil fertility, *Plant Soil,* Special vol., 423—442, 1971.
105. **Vancura, V., Abd-el-Malek, Y., and Zayed, M. N.,** *Azotobacter* and *Beijerinckia* in soil and rhizosphere of plant in Egypt., *Folia Microbiol. (Prague),* 10, 224—229, 1965.
106. **Derx, H. G.,** Free living heterotrophic nitrogen-fixing bacteria, in *The Biology of Nitrogen Fixation,* Quispel, A., Ed., North-Holland, Amsterdam, 1950, 37—85.
107. **Dobereiner, J., Day, J. M., and Dart, P. J.,** Nitrogenase activity and oxygen sensitivity of the *Paspalum notatum-Azotobacter paspali* association, *J. Gen. Microbiol.,* 71, 103—116, 1972.
108. **Brouzes, R., Mayfield, C. I., and Knowles, R.,** Effect of oxygen partial pressure on nitrogen fixation and acetylene reduction in a sandy loam soil amended with glucose, *Plant and Soil,* Special vol., 481—494, 1971.
109. **Jensen, H. L.,** Non-symbiotic nitrogen fixation, in *Soil Nitrogen,* Bartholomew, W. W. and Clark, F. E., Eds., *Am. Soc. Agron. Monograph,* 10, 436—480, 1965.
110. **Macura, J. and Kunc, F.,** Continuous flow method in soil microbiology. II. Observations on glucose metabolism, *Folia Microbiol. (Prague),* 6, 398—407, 1961.
111. **Roquerol, T.,** Sur le phenomene de fixation del'azote dans les rizieres de Camargue, *Ann. Agron. Paris,* 13, 325—46, 1962.
112. **Dobereiner, J., Day, J. M., and Dart, P. J.,** Nitrogenase activity in the rhizosphere of sugar cane and other tropical grasses, *Plant Soil,* 37, 191—196, 1972.
113. **Admase, A. D., Hoeks, J., De Bont, J. A. M., and Van Kessel, J. F.,** Microbial activities in soil near natural gas leaks, *Arch. Mikrobiol.,* 83, 32—51, 1972.
114. **Dobereiner, J. and Alvahydo, R.,** Sobre a influencia da Canada acucar na ocorrencia de "Beijerinckia" no solo. II. Influencia das diversas partes do vegetal, *Rev. Bras. Biol.,* 19, 401—12, 1959.
115. **Ruinen, J.,** The phyllosphere. I. An ecologically neglected milieu, *Plant Soil,* 15, 81—109, 1961.
116. **Ruinen, J.,** Nitrogen fixation in the phyllosphere, in *The Biology of Nitrogen Fixation,* Quispel, A., Ed., North-Holland, Amsterdam, 1974, 121—167.
117. **Dobereiner, J. and Campelo, A. B.,** Non-symbiotic nitrogenfixing bacteria in tropical soils, *Plant Soil,* Special vol., 557—570, 1971.
118. **Evans, H. J., Campbell, N. E. R., and Hill, S.,** Asymbiotic nitrogen-fixing bacteria from the surface of nodule and roots of legumes, *Can. J. Microbiol.,* 18, 13—21, 1972.
119. **Raju, P. N., Evans, H. J., and Seilder, R. J.,** An asymbiotic nitrogen-fixing bacterium from the root environment of corn, *Proc. Natl. Acad. Sci., USA,* 69, 3474—8, 1972.
120. **Bessems, E. P. M.,** Nitrogen Fixation in the Phyllosphere of Gramineae, Ph.D. thesis, University of Wageningen, Wageningen, Netherland, 1973.
121. **Mulder, E. G. and Veen, W. L. V.,** Effect of pH and organic compounds on nitrogen fixation by red clover, *Plant Soil,* 13, 91—113, 1960.
122. **Becking, J. H.,** Studies on nitrogen-fixing bacteria of the genus *Beijerinckia, Plant Soil,* 14, 49—81, 1961.
123. **Mulder, E. G. ad Brotonegoro, S.,** Free-living heterotrophic nitrogen-fixing bacteria, in *The Biology of Nitrogen Fixation,* Quispel, A., Ed., North-Holland, Amsterdam, 1974, 37—85.
124. **Tschapek, M. and Giambiagi, N.,** The formation of Leisegang's by *Azotobacter* due to O₂-inhibition, *Trans. 5th Int. Congr. Soil Sci.,* Leopoldville, Ill., 17, 97—103, 1954.

125. **Burns, R. C. and Hardy, R. W. F.,** *Nitrogen Fixation in Bacteria and Higher Plants,* Springer-Verlag, Basel, 1975.

126. **Ruinen, J.,** Nitrogen-fixation in the phyllosphere, in *Nitrogen Fixation by Free-Living Microorganisms,* Stewart, W. D. P., Ed., Cambridge University Press, New York, 1975, 85—100.

127. **Dobereiner, J.,** Nitrogen-fixing bacteria in the rhizosphere, in *The Biology of Nitrogen Fixation,* Quispel, A., Ed., North-Holland, Amsterdam, 1974, 86—120.

128. **Knowles, R.,** The significance of asymbiotic dinitrogen fixation by bacteria, in *Treatise on Dinitrogen Fixation,* Hardy, R. W. F. and Gibson, A. H., Eds., John Wiley & Sons, New York, 1977, 33—83.

129. **Leiser, A. T.,** A mucilaginous root sheath in Ericacea, *Am. J. Bot.,* 55, 391—398, 1968.

130. **Greaves, M. P. and Darbyshire, J. F.,** The ultrastructure of the mucilaginous layer on plant roots, *Soil Biol. Biochem.,* 4, 443—449, 1972.

131. **Foster, R. C. and Rovira, A. D.,** The rhizosphere of wheat roots studied by electron microscopy of ultrathin section, *Bull. Ecol. Res. Comm. (Stockholm),* 17, 93—102, 1978.

132. **Guckert, A., Breisch, H., and Reisinger, O.,** Interface soil-racine. I. Etude au microscope electronique des relations mucigelargil-microorganismes, *Soil Biol. Biochem.,* 7, 241—250, 1975.

133. **Gray, T. R. G.,** Stereoscan electron microscopy of soil microorganisms, *Science,* 155, 1668—1670, 1967.

134. **Marchant, R.,** The root surface of *Ammophila arenaria* as substrate for microorganisms, *Trans. Br. Mycol. Soc.,* 54, 479—482, 1970.

135. **Dart, P. J.,** Scanning electron microscopy of plant roots, *J. Exp. Bot.,* 22, 163—168, 1971.

136. **Campbell, R. and Rovira, A. D.,** The study of the rhizosphere by scanning electron microscopy, *Soil Biol. Biochem.,* 5, 747—752, 1973.

137. **Rovira, A. D. and Campbell, R.,** Scanning electron microscopy of microorganisms on the roots of wheat, *Microbial. Ecol.,* 1, 15—23, 1974.

138. **Old, K. M. and Nicholson, T. H.,** Electron microscopical studies of the microflora of roots of sand dune grasses, *New Phytol.,* 74, 51—58, 1975.

139. **Balandreu, J., Rinaudo, G., Fares-Hamed, I., and Dommergues, Y.,** Nitrogen fixation in the rhizosphere of rice plants, in *Nitrogen Fixation by Free-Living Microorganisms,* Stewart, W. D. P., Ed., Cambridge University Press, New York, 1975, 57—70.

140. **Samtsevich, A.,** Root excretions of plants. An important source of humus formation in the soil, *Humus Planta,* 5, 147—154, 1971.

141. **Dobereiner, J. and Day, J. M.,** Associative symbiosis in tropical grasses: characterization of microorganisms and dinitrogen fixing sites, *Proc. Int. Symp. N_2-Fixation,* Washington State Univ. Press, Pullman, Wash., 1975, 518—538.

142. **Von Bulow, J. W. F. and Dobereiner, J.,** Potential for nitrogen fixation in maize genotypes in Brazil, *Proc. Nat. Acad. Sci., USA,* 72, 2389—2393, 1975.

143. **Martin, J. P.,** Decomposition and binding action of polysaccharides in soil, *Soil Biol. Biochem.,* 3, 33—41, 1971.

144. **Knowles, R.,** Factors affecting dinitrogen fixation by bacteria in natural and agricultural systems, in *Proc. Int. Symp. N_2 Fixation,* Newton, W. E. and Nyman, C. J., Eds., Washington State Univ. Press, Pullman, Washington, 1976, 539—555.

145. **Hauke-Pacewiczowa, T., Balandreau, J., and Dommergues, Y.,** Fixation microbienne de l'azote dans un sol salin tunisien, *Soil Biol. Biochem,* 3, 47—53, 1970.

146. **Knowles, R. and Denike, D.,** Effect of ammonium, nitrite and nitrate-nitrogen on anaerobic nitrogenase activity in soil, *Soil Biol. Biochem.,* 6, 353—358, 1974.

147. **Patriquin, D. and Knowles, R.,** Nitrogen fixation in the rhizosphere of marine agiosperms, *Marine Biol.,* 16, 49—58, 1972.

148. **Dommergues, Y., Balandreau, J., Rinaudo, G., and Weinhard, P.,** Non-symbiotic nitrogen fixation in rhizospheres of rice, maize and different tropical grasses, *Soil Biol. Biochem.,* 5, 83—89, 1973.

149. **Day, J. M. and Dobereiner, J.,** Physiological aspects of N_2-fixation of *Sprillum* from *Digitaria* roots, *Soils Biol. Biochem.,* 7, 45—50, 1975.

150. **Patriquin, D. and Knowles, R.,** Effect of oxygen, mannitol and ammonium concentrations on nitrogenase activity in a marine skeletal carbonate sand, *Marine Biol.,* 32, 49—62, 1975.

151. **Greenwood, D. J.,** Studies on the distribution of oxygen around the roots of mustard seedlings (*Sinapis alba* L.), *New Phytol.,* 70, 97—101, 1971.

152. **Amstrong, W.,** Rhizosphere oxidation in rice; an analysis of inter-varietal differences in oxygen flux from the roots, *Physiol. Plant.,* 22, 296—303, 1969.

153. **Luxmoore, R. J., Stolzy, L. H., and Letey, J.,** Oxygen diffusion in the soil-plant system. IV. oxygen concentration profiles, respiration rates and radial oxygen losses predicted for rice roots, *Agron. J.,* 62, 329—332, 1970.

154. **Yoshida, T. and Ancajas, R.,** Nitrogen fixation by bacteria in the root zone of rice, *Soil Sci. Soc. Am. Proc.,* 35, 156—157, 1971.

155. **Teal, J. M. and Kanwisher, J. W.,** Gas transport in the marsh grass, *Spartina alternuiflora, J. Exp. Bot.,* 17, 355—361, 1966.

156. **Dalton, H. and Postage, J. R.,** Effect of oxygen on growth of *Azotobacter chroococcum* in batch and continuous cultures, *J. Gen. Microbiol.,* 54, 463—473, 1968.

157. **Greenwood, D. J. and Goodman, D.,** Direct measurements of the distribution of oxygen in soil aggregates and in columns of fine soil crumbs, *J. Soil. Sci.,* 18, 182—186, 1967.

158. **Mayfield, C. I. and Aldworth, R. L.,** Acetylene reduction by non-symbiotic bacteria in artificial soil aggregates amended with glucose, *Can J. Microbiol.,* 20, 877—881, 1974.

159. **Klucas, R.,** Nitrogen fixation by *Klebsiella* grown in the presence of oxygen, *Can. J. Microbiol.,* 18, 1845—1850, 1972.

160. **O'Toole, P. and Knowles, R.,** Efficiency of acetylene reduction (nitrogen fixation) in soil, effect of type and concentration of available carbohydrate, *Soil Biol. Biochem.,* 5, 789—797, 1973.

161. **Balandreau, J. and Fares-Hamad, I.,** Importance de la fixation d'azote dans la rhizosphere du riz, Colloque Rhizosphere, *Soc. Bot. France,* 122, 109—119, 1975.

162. **Bristow, M.,** Nitrogen fixation in the rhizosphere of freshwater angiosperms, *Can. J. Bot.,* 52, 217—221, 1974.

163. **Zuberer, D. A. and Silver, W. S.,** Mangrove-associated nitrogen fixation, in *Biology and Management of Mangroves,* Vol. 2., Walsh, G., Shedaker, S., and Teas, H. J., Eds., University Press, Gainesville, Fla., 1975, 643—653.

164. **Trinick, M. J.,** Host-*Rhizobium* associations, in *Nitrogen Fixation in Legumes,* Vincent, J. M., Ed., Academic Press, New York, 1982, 111—122.

165. **Schmidt, E. L.,** Legume symbiosis: ecology of the legume root bacteria, in *Interactions between Non-Pathogenic Soil Microorganisms and Plants,* Dommergues, Y. R. and Krupa, S. V., Eds., Elsevier, Amsterdam, 1978, 269—303.

166. **Vincent, J. M.,** Environmental factors in the fixation of nitrogen by the legume, in, *Soil Nitrogen,* Bartholomew, W. V. and Clark, F. E., Eds., Am. Soc. Agron., Madison, Wis., 1965, 384—435.

167. **Marshall, K. C.,** Survival of root nodule bacteria in dry soils exposed to high temperatures, *Aust. J. Agric. Res.,* 15, 273—281, 1964.

168. **Gibson, A. H.,** Physical environment and symbiotic nitrogen fixation. IV. Factors affecting the early stages of nodulation, *Aust. J. Biol. Sci.,* 20, 1087—1104, 1967.

169. **Weber, D. F. and Miller, V. L.,** Effect of soil temperature of *Rhizobium japonicum* serogroup distribution in soybean nodules, *Agron. J.,* 64, 796—798, 1972.

170. **Ek-Jander, J. and Fahraeus, G.,** Adaption of *Rhizobium* to sub-arctic environment in Scandinavia, *Plant Soil,* Spec. vol., 129—137, 1971.

171. **Masterson, C. L. and Sherwood, M. T.,** Selection of *Rhizobium trifolii* strains by white and subterranean Clovers, *Ir. J. Agric. Res.,* 13, 91—99, 1974.

172. **Foulds, W.,** Effect of drought on three species of *Rhizobium, Plant Soil,* 35, 665—667, 1971.

173. **Hamdi, Y. A.,** Soil Water tension and the movement of rhizobia, *Soil Biol. Biochem.,* 3, 121—136, 1971.

174. **Holding, A. J. and King, J.,** The effectiveness of indigenous populations of *Rhizobium trifolii* in relation to soil factors, *Plant Soil,* 18, 191—198, 1963.

175. **Means, U. M., Johnson, H. W., and Date, R. A.,** Quick Serological method of classifying strains of *Rhizobium japonicum* in nodules, *J. Bacteriol.,* 87, 547—553, 1964.

176. **Damirgi, S. M., Frederick, L. R., and Anderson, I. C.,** Serogroups or *Rhizobium japonicum* in nodules as affected by soil types, *Agron. J.,* 59, 10—12, 1967.

177. **Ham, G. E., Fredrick, L. R., and Anderson, I. C.,** Serogroups of *Rhizobium japonicum* in soybean nodules as affected by soil types, *Agron. J.,* 59, 10—12, 1967.

178. **Holland, A. A. and Parker, C. A.,** Studies on microbial antagonism in the establishment of clover pasture. II. The effect of saprophytic soil fungi upon *Rhizobium trifolii* and the growth of subterranean clover, *Plant Soil,* 25, 329—340, 1966.

179. **Parker, C. A. and Grove, P. L.,** *Bdellovibrio bacteriovortus* parasitizing *Rhizobia* in Western Australia, *J. Appl. Bacteriol.,* 33, 253—255, 1970.

180. **Kowalski, M., Ham, G. E., Frederick, L. R., and Anderson, I. C.,** Relationship between strains of *Rhizobium japonicum* and their bacteriophages from soil and nodules of field-grown soybeans, *Soil Sci.,* 118,, 221—228, 1974.

181. **Roslycky, E. B.,** Bacteriocin production in the rhizobia bacteria, *Can. J. Microbiol.,* 13, 341—432, 1967.

182. **Lotz, W. and Mayer, F.,** Isolation and characterization of bacteriophage tail-like bacteriocin from a strain of *Rhizobium, J. Virol.,* 9, 160—173, 1972.

183. **Schwinghamer, E. A., Pankhurst, C. E., and Whitfield, P. R.,** A phage like bacteriocin of *Rhizobium trifolii, Can. J. Microbiol.,* 19, 359—368, 1973.

184. **Schwinghamer, E. A. and Dudman, W. R.,** Evaluation of spectinomycin resistance as a marker for ecological studies with *Rhizobium* spp., *J. Appl. Bacteriol.,* 36, 263—272, 1973.

185. **Clayet-Marel, J. C. and Crozat, Y.**, Etude ecologique en immunofluorescence de *Rhizobium japonicum* dans le sol et al rhizosphere, *Agronomie*, 2, 243—248, 1982.

186. **Moawad, H. A., Ellis, W. R., and Schmidt, E. L.**, Rhizosphere response as a factor in competition among three serogroups of indigenous *Rhizobium japonicum* for nodulation of field grown soybean, *Appl. Environ. Microbiol.*, 47, 607—612, 1984.

187. **Robert, F. M. and Schmidt, E. L.**, Population changes and persistence of *Rhizobium phaseoli* in soil and rhizospheres, *Appl. Environ. Microbiol.*, 45, 550—556, 1983.

188. **Reyes, V. G. and Schmidt, E. L.**, Population densities of *Rhizobium japonicum* strain 123 estimated directly in soil and rhizospheres, *Appl. Environ. Microbiol.*, 37, 854—858, 1979.

189. **Reyes, V. G. and Schmidt, E. L.**, Populations of *Rhizobium Japonicum* associated with the surfaces of soil - grown roots, *Plant Soil*, 61, 71—80, 1981.

190. **Dudman, W. F. and Brockwell, J.**, Ecological studies of root nodule bacteria introduced into field environments. I. A survey of field performance of clover inoculants by gel immune diffusion serology, *Aust. J. Agric. Res.*, 19, 739—747, 1968.

191. **Chatel, D. L. and Parker, C. A.**, Survival of field-grown rhizobia over the dry summer period in western Australia, *Soil Biol. Biochem.*, 5, 415—423, 1973.

192. **Chatel, D. L., Shipton, W. A., and Parker, C. A.**, Establishment and persistence of *Rhizobium trifolii* in Western Australia soils, *Soil Biol. Biochem.*, 5, 815—824, 1973.

193. **Chatel, D. L. and Greenwood, R. M.**, The location and distribution in soil of rhizobia under sensesced annual legume pastures, *Soil Biol. Biochem.*, 5, 799—808, 1973.

194. **Bohlool, B. B. and Schmidt, E. L.**, A fluorescent antibody technique determination of growth rates of bacteria in soil, in *Modern Methods in the Study of the Microbial Ecology*, Rosswall, T., Ed., *Bull. Eco. Res. Comm. (Stockholm)*, 17, 336—338, 1973.

195. **Schmidt, E. L.**, Quantitative autecological study of microorganisms in soil by immunofluorescence, *Soil Sci.*, 118, 141—149, 1974.

196. **Rovira, A. D. and Stern, W. R.**, The rhizosphere bacteria in grass clover associations, *Aust. J. Agric. Res.*, 12, 1108—1118, 1961.

197. **Katzelson, H.**, Nature and importance of rhizosphere, in *Ecology of Soil-Borne Plant Pathogens*, Baker, K. F. and Snyder, W. C., Eds., University of California Press, Berkeley, Calif., 1965, 187—209.

198. **Tuzimura, K. and Watanabe, I.**, The growth of *Rhizobium* in the rhizosphere of the host plant. Ecological studies of root nodule bacteria, *Soil Sci. Plant Nutr.*, 8, 10—24, 1961.

199. **Rovira, A. D.**, *Rhizobium* numbers in the rhizosphere of red clover and *Paspalum* in relation to soil treatment and the numbers of bacteria and fungi, *Aust. J. Agric. Res.*, 12, 77—83, 1961.

200. **Tuzimura, K. and Watanabe, I.**, The effect of rhizosphere of various plants on the growth of *Rhizobium*. Ecological studies on root nodule bacteria, *Soil. Sci. Plant Nutr.*, 8, 13—17, 1962.

201. **Nutman, P. S.**, The relation between root hair infection by *Rhizobium* and nodulation in *Trifolium* and *Vicia*, *Proc. R. Soc. London, Ser. B.*, 156, 122—137, 1962.

202. **Krasil'nikov, A. A.**, *Soil Microorganisms and Higher Plants*, Acad. Sci., USSR, (English edition published for the National Science Foundation and Department of Agriculture, USA) 1958.

203. **Nutman, P. S.**, The relation between nodule bacteria and legume host in the rhizosphere and in the process of infection, in *Ecology of Soil Borne Plant Pathogens*, Baker, K. F. and Snyder, W. C., Eds., University of California Press, Berkeley, Calif., 1965, 231—247.

204. **Brown, M. E., Jackson, R. M., and Burlingham, S. K.**, Growth and effects of bacteria introduced into soil, in *The Ecology of Soil Bacteria*, Gray, T. R. G. and Perkinson, D., Eds., Liverpool University Press, London, 1968, 531—551.

205. **Robinson, A. C.**, The influence of host on soil and rhizosphere populations of clover and lucerne nodule bacteria in the field, *J. Aust. Inst. Agric. Sci.*, 33, 207—209, 1967.

206. **Diatloff, A.**, The introduction of *Rhizobium japonicum* to soil by seed inoculation of non-host legumes and cereals, *Aust. J. Exp. Agric. Anim. Husb.*, 9, 357—360, 1969.

207. **Kvien, C. S., Ham, G. E., and Lambert, J. W.**, Recovery of introduced *Rhizobium japonicum* strains by soybean genotypes, *Agron. J.*, 73, 900—905, 1981.

208. **Vest, G., Weber, D. F., and Sloger, C.**, Nodulation and nitrogen fixation, in *Soybeans: Improvement, Production, and Uses*, Coldwell, B. E., Ed., Soc. Agron., Madison, Wis., 1973, 353—390.

209. **Weaver, R. W. and Frederick, L. R.**, Effect of inoculum rate of competitive nodulation of *Glycine max* L. Merril. II. Field studies, *Agron. J.*, 66, 233—236, 1974.

210. **Kapusta, G. and Rouwenhorst, D. L.**, Interaction of selected pesticides on *Rhizobium japonicum* in pure culture under field conditions, *Agron. J.*, 65, 112—115, 1973.

211. **Bohlool, B. B. and Schmidt, E. L.**, Persistence and competition aspects of *Rhizobium japonicum* observed in soil by immunofluorescence microscopy, *Soil Sci. Soc. Am. Proc.*, 37, 561—564, 1973.

212. **Ham, G. E., Cardwell, V. B., and Johnson, H. W.**, Evaluation of *Rhizobium japonicum* inoculants in soils containing naturalized populations of rhizobia, *Agron. J.*, 63, 301—303, 1971.

213. **Egeraat, A. W. and Van, S. M.,** Pea root exudates and their effect upon root nodule bacteria, *Meded. Landbouwhogesck. Wageningen,* H. Veenman en Zonen N. V., Wageningen, The Netherlands, 1972, 90.
214. **Dart, P. J. and Marcer, F. V.,** The legume rhozosphere, *Arch. Mikrobiol.,* 47, 344—378, 1964.
215. **Fahraeus, G. and Ljunggren, H.,** Pre-infection phases of the legume symbiosis, in *The Ecology of Soil Bacteria,* Gray, T. R. G. and Perkinson, D., Eds., Liverpool, London, 1968, 396—421.
216. **Solpheim, B. and Raa, J.,** Evidence countering the theory of specific induction of pectin-degrading enzymes as basis for specificity in *Rhizobium*-Leguminosae association, *Plant Soil,* 35, 275—280, 1971.
217. **Hubbell, D. H.,** Studies on the root hair "curling factor" of Rhizobium, *Bot. Gaz.,* 131, 337—342, 1970.
218. **Bohlool, B. B. and Schmidt, E. L. M.,** Lectins a possible basis for specificity in the *Rhizobium legume* root nodule symbiosis, *Science,* 185, 269—217, 1974.
219. **Hamlin, J. and Kent, S. P.,** Possible role of phytohaemagglutinin in Phaseolus vulgaris L., *Nature New Biol.,* 245, 28—30, 1973.
220. **Sharon, N. and Lis, H.,** Lectins cell agglutinating and sugar-specific proteins, *Science,* 177, 949—959, 1972.
221. **Schmidt, E. L., Bankole, R. O., and Bohlool, B. B.,** Fluorescent antibody approach to study rhizobia in soil, *J. Bacteriol.,* 95, 1987—1992, 1968.
222. **Sahlman, K. and Fahraeus, G.,** An electron microscope study of root hair infection by Rhizobium, *J. Gen. Microbiol.,* 33, 425—427, 1963.
223. **Nutman, P. S.,** Perspectives in biological nitrogen-fixation, *Sci. Prog.,* 59, 55—74, 1971.
224. **Fisher, D. J., Hayes, A. L., and Jones, C. A.,** Effects of some surfactant fungicides on *Rhizobium trifolii* and its symbiotic relationship, *Ann. Appl. Biol.,* 90, 73—84, 1978.
225. **Brockwell, J., Diatloft, A., Roughley, R. J., and Date, R. A.,** selection of *Rhizobia* for inoculants, in *Nitrogen Fixation in Legumes,* Vincent J. M., Ed., Academic Press, New York, 1982, 173—191.
226. **Vincent, J. M.,** Factors controlling the legume-*Rhizobium* symbiosis, in *Nitrogen fixation,* Vol II., Newton, W. E. and Orme-Johnson, W. H., Eds., University Park Press, Baltimore, 1980, 103—139.
227. **Badenoch-Jones, J., Flanders, D. J., and Rolfe, B. G.,** Association of *Rhizobium* strains with roots of *Trifolium repens, Appl. Environ. Microbiol.,* 49, 1511—1520, 1985.
228. **Li, D. and Hubbell, D. H.,** Infection thread information as basis of nodulation specificity in *Rhizobium* strawberry clover association, *Can. J. Microbiol.,* 15, 1133—1136, 1969.
229. **Napoli, C. A. and Hubbell, D. H.,** Ultrastructure of *Rhizobium* induced infection threads in clover root hairs, *Appl. Microbiol.,* 30, 1003—1009, 1975.
230. **Yao, P. Y. and Vincent, J. M.,** Host specificity in the root hair, "curling factor" of *Rhizobium* sp., *Aust. J. Biol. Sci.,* 22, 413—423, 1969.
231. **Dazzo, F. B. and Hubbell, D. H.,** Cross reactive analysis and lectin as determinants of symbiotic specificity in the *Rhizobium*-clover association, *Appl. Environ. Microbiol.,* 30, 1017—1033, 1975.
232. **Dazzo, F. B., Urbano, M. R., and Bril, W. J.,** Transient appearance of lectin receptors of *Rhizobium trifolii, Curr. Microbiol.,* 2, 15—20, 1979.
233. **Dazzo, F. B., Brabak, E. M., Urbano, M. R., Sherwood, J. E., and Truchet, G. L.,** Regulation of recognition in *Rhizobium* clover symbiosis, in *Current Perspectives in Nitrogen Fixation,* Gibson, A. H. and Newton, W. E., Eds., Australian Academy of Science, Canberra, Australia, 1981, 252—295.
234. **Dazzo, F. B., Truchet, G. L., and Kijne, J. W.,** Lectin involvement in root-hair tip adhesion as related to the *Rhizobium* clover symbiosis, *Physiol. Plant,* 56, 143—147, 1982.
235. **Elkan, G. H.,** The taxonomy of Rhizobiaceae, in *Biology of the Rhizobiaceae,* Giles, K. and Atherly, A., Eds., Academic Press, New York, 1981, 1—14.
236. **Bauer, W. D.,** Infection of legumes by Rhizobia, *Annu. Rev. Plant Physiol.,* 32, 407—449, 1981.
237. **New Comb, W.,** Nodule morphogenesis and differentiation, in *Biology of Rhizobiaceae,* Giles, K. and Atherly, A., Eds., Academic Press, New York, 1981, 247—298.
238. **Banfalvi, Z., Sakanyan, V., Koncz, C., Kiss, A., Dusha, I., and Kondorosi, A.,** Localization of nodulation and nitrogen fixation genes on a high molecular weight plasmid of *Rhizobium meliloti, Mol. Gen. Genet.,* 184, 318—324, 1981.
239. **Downie, J. A., Hombrecher, G., Ma, Q. S., Knight, C., Wells, B., and Johnson, A. W. B.,** Cloned nodulation genes of *Rhizobium leguminosarum* determine host range specificity, *Mol. Gen. Genet.,* 190, 359—365, 1983.
240. **Kondorosi, E., Banfalvi, Z., and Kondorosi, A.,** Physical and genetic analysis of a symbiotic region of *Rhizobium meliloti.* Identification of nodulation genes, *Mol. Gen. Genet.,* 193, 445—452, 1982.
241. **Rushenberg, C., Biostard, P., Denarie, J., and Casse-Debart, F.,** Genes controlling early and late function in symbiosis are located on a meger plasmid in *Rhizobium meliloti, Mol. Gen. Genet.,* 1984, 326—333, 1981.
242. **Schofield, P. R., Djordjevic, M. A., Rolfe, B. G., Shine, J., and Watson, J. M.,** Molecular linkage map of nitrogenase and nodulation genes in *Rhizobium trifolii, Mol. Gen. Genet.,* 192, 459—465, 1983.
243. **Fisher, R. F., Tu, J. K., and Long, S. R.,** Conserved nodulation genes in *Rhizobium meliloti* and *Rhizobium trifolii, Appl. Environ. Microbiol.,* 49, 1432—1435, 1985.

244. **Jacobs, T. W., Egelhoff, T. T., and Long, S. R.,** Physical and genetic map of a *Rhizobium meliloti* nodulation gene region and nucleotide sequence of nod C, *J. Bacteriol.,* 162, 469—476, 1985.

245. **Djordijevic, M. A., Zurkowski, W., Shine, J., and Rolfe, B. G.,** Sym plasmid transfer to various symbiotic mutants of *Rhizobium trifolii, Rhizobium leguminosarum* and *Rhizobium meliloti, J. Bacteriol.,* 156, 1035—1045, 1983.

246. **Hombrecher, G., Gotz, R., Dibb, N. J., Downie, J. A., Johnson, A. W. B., and Brewin, N. J.,** Cloning and mutagenesis of nodulation genes from *Rhizobium leguminosarum* TOM, a strain with extended host range, *Mol. Gen. Genet.,* 194, 293—298, 1984.

247. **Adams, M. W. and Pipoly, J. J.,** Biological structures classification and distribution of economic legumes, in *Advances in Legume Science,* Royal Botanical Gardens, Kew, England, 1979, 1—16.

248. **Fleming, A. A., Giddens, J. E., and Beaty, E. R.,** Corn yield as related to legumes and inorganic nitrogen, *Crop Sci.,* 21, 977—980, 1981.

249. **Nelson, D. R., Bellville, R. J., and Porter, C. A.,** Role of nitrogen assimilation in seed development of soybean, *Plant Physiol.,* 74, 128—133, 1984.

250. **Gibson, A. H.,** Factors in the physical and biological environment affecting nodulation and nitrogen fixation in legumes, *Plant Soil,* Special vol., 139—152, 1971.

251. **Loneragan, J. F.,** The soil chemical environment in relation to symbiotic nitrogen fixation, in *Use of Isotopes for Study of Fertilizer Utilization by Legume Crops,* International Atomic Energy Agency, Vienna, 1972, No. 149, 17—54.

252. **Nutman, P. S.,** The influence of physical environmental factors on the activity of *Rhizobium,* in *Soil and Symbiosis in Use of Isotopes for Study of Fertilizer Utilization by Legume Crops,* International Atomic Energy Agency, Vienna, 1972, No. 149, 55—84.

253. **Pate, J. S.,** Physiology of the reaction of nodulated legumes to environment, in *Symbiotic Nitrogen Fixation in Plants,* Nutman, P. S., Ed., Cambridge University Press, New York, 1976, 335—360.

254. **Gibson, A. H.,** Recovery and compensation by nodulated legumes to environmental stresses, in *Symbiotic Nitrogen Fixation in Plants,* Nutman, P. S., Ed., Cambridge University Press, New York, 1976, 384—404.

255. **Schlegel, H. G.,** Die Isolierung von Poly-β-hydroxy-bettersaure aus Wurzelknollchen von Leguminosen, *Flora,* 152, 236—240, 1962.

256. **Wong, P. P. and Evans, H. J.,** Poly-β-hydroxybutyrate utilization by soybeans (*Glycine max* Merr.) nodules and assessment of its role in maintenance of nitrogenase activity, *Plant Physiol.,* 47, 750—755, 1971.

257. **Minchin, F. R. and Pate, J. S.,** Diurnal functioning of the legume root nodule, 25, 295—308, 1974.

258. **Jones, R. J., Davies, J. G., and Waite, R. B.,** The contribution of some tropical legumes to pasture yield of dry matter and nitrogen at Sandford, SE Queensland, *Aust. Exp. Agri. Anim. Husb.,* 7, 57—65, 1967.

259. **Hardy, R. W. F., Holsten, R.D., Jackson, E. K., and Burns, R. C.,** The acetylene-ethylene assay for N_2 fixation: laboratory and field evaluations, *Plant Physiol.,* 43, 1185—1207, 1968.

260. **Sprent, J. I.,** Nitrogen fixation by legumes subjected to water and light stresses, in *Symbiotic Nitrogen Fixation in Plants,* Nutman, P. S., Ed., Cambridge University Press, New York, 1976, 405—420.

261. **Hardy, R. W. F. and Havelka, V. D.,** Symbiotic N_2 fixation: multi-fold enhancement of CO_2-enrichment of field-grown soybean, *Plant Physiol.,* 35, 1973.

262. **Streeter, J. G.,** Growth of two shoots on a single root as a technique for studying physiological factors limiting the rate of nitrogen fixation by nodulated legumes, *Plant Physiol.,* 35, 1973.

263. **Carr, D. J. and Pate, J. S.,** Ageing in the whole plant, *Sym. Soc. Exper. Biol.,* 21, 559—600, 1967.

264. **Roponen, I. E. and Virtanen, A. I.,** The effect of prevention of flowering on the vegetative regrowth of inoculated pea plants, *Physiol. Plant.,* 21, 655—677, 1968.

265. **Hardy, R. W. F., Burns, R. C., Herbert, R. R., Holsten, R. D., and Jackson, E. K.,** Biological nitrogen fixation: a key to world protein, *Plant Soil,* Special vol., 561—590, 1971.

266. **Pate, J. S. and Flinn, A. M.,** Carbon and nitrogen transfer from vegetative organs to ripening seeds of field pea (*Pisum arvense* L.), *Exp. Bot.,* 24, 1090—1099, 1973.

267. **Virtanen, A. I., Moisio, T., and Burris, R. H.,** Fixation of nitrogen by nodules excised from illuminated and darkened pea plants, *Acta Chem. Scand.,* 9, 184-186, 1955.

268. **Bergersen, F. J.,** The quantitative relationship between nitrogen fixation and the acetylene reduction assay, *Aust. Biol.,* 23, 1015—1025, 1970.

269. **Mague, T. H. and Burris, R. H.,** Reduction of acetylene and nitrogen by field-grown soybeans, *New Phytolog.,* 71, 275—86.

270. **Day, J. M.,** Studies of the Role of Light Legume Symbiosis, Ph.D. thesis, University of London, 1972.

271. **Roponen, I. E., Valle, E., and Ettala, T.,** Effect of temperature of the culture medium on growth and nitrogen fixation in inoculated legumes and rhizobia, *Physiol. Plant.,* 23, 1198—1205, 1970.

272. **Pate, J. S.,** *Physiological Aspects of Inorganic and Intermediate Nitrogen Metabolism,* Academic Press, New York, 1968.
273. **Pate, J. S., Gunning, B. E. S., and Bariarty, L. G.,** Ultrastructure and functioning of the transport system of the leguminous root nodule, *Planta,* 85, 11—34, 1969.
274. **Minchin, F. R., and Pate, J. S.,** The carbon balance of a legume and the functional economy of its root nodules, *J. Exp. Bot.,* 24, 259—271, 1973.
275. **Pate, J. S.,** Movement of nitrogenous solutes in plants, in *Nitrogen in Soil-Plant Studies,* International Atomic Energy Agency, Vienna, 1971.
276. **Pate, J. S.,** Uptake, assimilation and transport of nitrogen compounds by plants, *Soil Biol. Biochem.,* 5, 109—119, 1973.
277. **Small, J. G. C. and Leonard, D. A.,** Translocation of ^{14}C-labelled photosynthate in nodulated legumes as influenced by nitrate nitrogen, *Am. J. Bot.,* 56, 187—196, 1969.
278. **Oghoghorie, C. G. O. and Pate, J. S.,** The nitrate stress syndrome of the nodulated field pea (*Pisum arvense* L.), Techniques for measurement and evaluation in physiological terms, *Plant Soil,* Special vol., 185—202, 1971.
279. **Olsen, R. A.,** Fertilizing the soybean crop (*Glycine max* L.), in *Use of Isotopes for Study of Fertilizer Utilization by Legume Crops,* IAEA Publication No. 149, International Atomic Energy Agency, Vienna, 1972, 101—110.
280. **Sprent, J. I.,** The effects of water stress on nitrogen-fixing root nodules. I. Effects on the physiology of detached soybean nodules, *New Phytol.,* 70, 9—17, 1971.
281. **Sprent, J. I.,** The effects of water stress on nitrogen fixing root nodules. II. Effects on the fine structure of detached soybean nodules, *New Phytol.,* 71, 443—450, 1972.
282. **Sprent, J. I.,** The effects of water stress on nitrogen-fixing root nodules. III. Effects of osmotically applied stress, *New Phytol.,* 71, 451—460, 1972.
283. **Sprent, J. I.,** The effects of water stress on nitrogen-fixing root nodules. IV. Effects on whole plants of *Vicia faba* and *Glycine max., New Phytol.,* 71, 603—611, 1972.
284. **Engin, M. and Sprent, J. I.,** Effects of water stress on growth and nitrogen-fixing activity of *Trifolium repens, New Physiol.,* 72, 117—126, 1973.
285. **Ferguson, T. P. and Bond, G.,** Symbiosis of leguminous plants and nodule bacteria. V. The growth of red clover at different oxygen tensions, *Ann. Bot.,* 18, 385—396, 1954.
286. **Michin, F. R.,** Physiological Functioning of the Plant Nodule Symbiotic System of Garden Pea *(Pisum sativum)* cv. Meteor, Ph.D. thesis, Queen's University, Belfast, 1973.
287. **Bergerson, F. J.,** The biochemistry of symbiotic nitrogen fixation in legumes, *Ann. Rev. Plant Physiol.,* 22, 121—140, 1971.
288. **McManmon, M. and Crawford, R. M. M.,** A metabolic theory of flooding tolerance; the significance of enzyme distribution and behaviour., *New Phytol.,* 70, 299—306, 1971.
289. **Devitt, A. C. and Francis, C. M.,** The effect of waterlogging on the mineral nutrient content of *Trifolium subterraneum, Aust. J. Exp. Agric. Anim. Husb.,* 12, 614—617, 1972.
290. **Grobbelaar, N., Clarke, B., and Hough, M. C.,** The nodulation and nitrogen fixation of isolated roots of *Phaseolus vulgaris* L. III. Effects of carbon dioxide and ethylene, *Plant Soil,* Special vol., 216—223, 1971.
291. **Gibson, A. H.,** Physical environment and symbiotic nitrogen fixation VI. Nitrogen retention within the nodules of *Trifolium subterraneum* L., *Aust. J. Biol. Sci.,* 22, 829—838, 1969.
292. **Gibson, A. H.,** Physical environment and symbiotic nitrogen fixation. VII. Effect of fluctuating root temperature on nitrogen fixation, *Aust. J. Biol. Sci.,* 22, 839—846, 1969.
293. **Davidson, J. L., Gibson, A. H., and Birch., J. W.,** Effects of temperature and defoliation on growth and nitrogen fixation in subterranean clover, *Proc. XI Inter. Cong.* University of Queensland Press, Surfer's Paradise, Australia, 1970, 542—545.
294. **Pankhurst, C. E. and Gibson, A. H.,** *Rhizobium* strain influence on disruption of clover nodule development at high root temperature, *J. Gen. Microbiol.,* 74, 219—231, 1973.
295. **De-Polli, H., Boyer, C. D., and Neyra, C. A.,** Nitrogenase activity associated with roots and stems of field grown corn (*Zea mayes*) plants, *Plant Physiol.,* 70, 1009—1013, 1982.
296. **Reynders, L. and Vlassak, K.,** Use of *Azospirillum brasilense* as biofertilizer in intensive wheat cropping, *Plant Soil,* 66, 217—273, 1982.
297. **Haahtela, K., Wartiovaara, T., Sundman, V., and Skujins, J.,** Root associated N_2-fixation (C_2H_2-reduction) by enterobacteriacea and *Azospirillum* strains in cold climate spodosols, *Appl. Environ. Microbiol.,* 41, 203—206, 1981.
298. **Dobereiner, J.,** Dinitrogen fixation in rhizosphere and phyllosphere associations, in *Encyclopedia of Plant Physiology,* Lauchli, A. and Bieleski, R., Eds., Vol. 15A, Springer, Basel, 1983, 330—350.
299. **Van Berkum, P. and Bohlool, B. B.,** Evaluation of nitrogen fixation by bacteria in association with roots of tropical grasses, *Microbiol. Rev.,* 44, 491—517, 1980.

300. **Berg, R. H., Tyler, M. E., Novick, N. J., Vasil, V., and Vasil, J. K.,** Biology of *Azospirillum* sugarcane association: enhancement of nitrogenase activity, *Appl. Environ. Microbiol.,* 39, 642—649, 1980.

301. **Schank, S. C., Weiner, K. L., and McRae, I. C.,** Plant yield and nitrogen content of a digitgrass in response to *Azospirillum* inoculation, *Appl. Environ. Microbiol.,* 39, 642—649, 1980.

302. **Okon, Y., Heytler, P. G., and Hardy, R. W. F.,** N_2-fixation by *Azospirillum brasilense* and its incorporation into host *Setaria italica, Appl. Environ. Microbiol.,* 46, 694—697, 1983.

303. **Neuer, G., Kronenberg, A., and Bothe, H.,** Denitrification and nitrogen fixation by *Azospirillum.* III. Properties of a wheat *Azospirillum* association, *Arch. Microbiol.,* 14, 364-370, 1983.

304. **Hartman, A., Singh, A., and Klingmuller, W.,** Isolation and characterisation of *Azospirillum* mutants excreting high amounts of indoleacetic acid, *Can. J. Microbiol.,* 29, 916—923, 1983.

305. **Lin, W., Okon, Y., and Hardy, R. W. F.,** Enhancement mineral of uptake by *Zea mays* and *Sorgham bicolor* roots inoculated with *Azospirillum brasilense, Appl. Environ. Microbiol.,* 45, 1775—1779, 1983.

306. **Smith, R. L., Schank, S. C., Milan, J. R.,and Baltensperger, A. A.,** Response of *Sorghum* and *Penisetum* species to N_2 fixing bacterium *Azospirillum brasilense, Appl. Environ. Microbiol.,* 47, 1331—1336, 1984.

307. **Nelson, M. and Knowles, R.,** Effect of oxygen and nitrate on nitrogen fixation and denitrification by *Azospirillum brasilense* grown in continuous culture, *Can. J. Microbiol.,* 24, 1395—1403, 1978.

308. **Tibelius, K. N. and Knowles, R.,** Uptake and hydrogenase activity in denitrifying *Azospirillum brasilense* grown anaerobically with nitrous oxide or nitrate, *J. Bacteriol.,* 157, 84—88, 1984.

309. **Bothe, H., Klein, B., Penteado, S. M., and Dobereiner, J.,** Transformation of inorganic nitrogen by *Azospirillum* spp., *Arch. Microbiol.,* 130, 96—100, 1981.

310. **Zimmer, W., Penteado, S. M., and Bothe, H.,** Denitrification by *Azospirillum brasilense* sp. 7. I. Growth with nitrite as respiratory electron acceptor, *Arch. Microbiol.,* 138, 206—211, 1984.

311. **Penteado, S. M., Zimmer, W., and Bothe, H.,** Denitrification *Azospirillum brasilense* sp. II. Growth with oxide as respiratory electron acceptor, *Arch. Microbiol.,* 138, 212—216, 1984.

312. **Stewart, W. D. P., Haystead, A., and Dharamawardence, M. W. N.,** Nitrogen assimilation and metabolism in blue green algae, in *Nitrogen Fixation by Free-living Microorganisms,* Stewart, W. D. P., Ed., Cambridge University Press, London, 1975, 129—158.

313. **Fogg, G. E.,** *Algal Culture and Phytoplankton Ecology,* 2nd ed., University of Wisconsin Press, Madison, 1975.

314. **Fogg, G. E., Stewart, W. D. P., Fay, P., and Walsby, A. E.,** *The Blue Green Algae,* Academic Press, New York, 1975.

315. **Carr, N. G. and Whitton, B. A.,** *The Biology of Blue Green Algae,* Blackwell Scientific, Oxford, 1973.

316. **Venkataraman, G. S.,** *Algal Biofertilizers and Rice Cultivation,* Today and Tomorrow's Printers, New Delhi, 1972.

317. **Wolk, C. P.,** Physiology and cytological chemistry of blue green algae, *Bacteriol. Rev.,* 36, 1—32, 1973.

318. **Henrikson, E.,** Algal nitrogen fixation in temperate regions, *Plant Soil,* Special vol., 415—419, 1971.

319. **Wiebe, W. J., Johnnes, R. E., and Webb, K. L.,** Nitrogen fixation in a coral reef community, *Science,* 188, 257—259, 1975.

320. **Hill, I. R. and Wright, S. J. L.,** *Pesticide Microbiology,* Academic Press, New York, 1978.

321. **Stewart, W. D. P., Haystead, A., and Pearson, H. W.,** Nitrogenase activity in heterocysts of blue-green algae, *Nature (London),* 224, 226—228, 1969.

322. **Stewart, W. D. P.,** *The Biology of Nitrogen Fixing Organisms,* McGraw-Hill, New York, 1980.

323. **Fogg, G. E.,** Nitrogen fixation by photosynthetic organisms, *Ann. Rev. Plant Physiol.,* 7, 51—70, 1956.

324. **Holm-Hansen, O.,** Ecology, physiology and biochemistry of blue-green algae, *Ann. Rev. Microbiol.,* 22, 47-70, 1968.

325. **Shtina, E. A.,** The distribution and role of nitrogen-fixing blue-green algae in soils of the temperate zone of the USSR, Trans. 9th Inter. Cong. Soil Sci., II, Adelaide, Australia, 1968.

326. **Jurgensen, M. F.,** Relationship between non-symbiotic nitrogen fixation and soil nutrients status. A review, *J. Soil Sci.,* 24, 512—522, 1973.

327. **Mishustin, E. N., Kaliniskaya, T. A., and Shemakhanova, N. M.,** Fixation of molecular nitrogen by microorganisms: review of works done in the USSR from 1966—1972, *Izd. Akad. Nauk. SSR. Ser. Biol.,* 6, 779—796, 1973.

328. **Stewart, W. D. P.,** Algal fixation of atmospheric nitrogen, *Plant Soil,* 32, 555—588, 1970.

329. **Stewart, W. D. P.,** Nitrogen fixation by photosynthetic microorganisms, *Ann. Rev. Microbiol.,* 27, 283—316, 1973.

330. **Stewart, W. D. P.,** Nitrogen fixation: differentiating cyanobacteria rearrange their nif genes, *Nature (London),* 314, 404—405, 1985.

331. **Stewart, W. D. P.,** Nitrogen turnover in marine and brackish habitats. II. Use of ^{15}N in measuring nitrogen fixation in the field, *Ann. Bot.,* 31, 385—407, 1967.

332. **Stewart, W. D. P. and Alexander, G.,** Phosphorus availability and nitrogenase activity in aquatic blue-green algae, *Freshwater Biol.,* 1, 389—404, 1970.

333. **Stewart, W. D. P. and Pearson, H. W.**, Effects of aerobic and anaerobic conditions on growth and metabolism of blue-green algae, *Proc. R. Soc. London*, 175, 293—311, 1970.

334. **Dalton, H.**, Fixation of dinitrogen by free-living microorganisms, *CRC Crit. Rev. Microbiol.*, 3, 183—220, 1974.

335. **Quispel, A.**, *The Biology of Nitrogen Fixation*, American Elsevier, New York, 1974.

336. **Pankratova, E. M.**, The participation of nitrogen-fixing algae in the accumulation of nitrogen in soil, *Izv. Akad. Nauk. SSSR. Ser. Biol.*, 2, 188—197, 1979.

337. **Mal'tseva, N. M.**, Role of free-living nitrogen-fixing organisms in the nitrogen balance in soil, *Microbiol. Zurn. (Kiev)*, 39, 780—781, 1977.

338. **Peters, G. A.**, Blue-green algae and algal associations, *Bioscience*, 28, 580—585, 1978.

339. **De, P. K. and Mandal, L. N.**, Fixation of nitrogen by algae in rice soils, *Soil Sci.*, 31, 453—458. 1956.

340. **De, P. K. and Sulaiman, M.**, Fixation of nitrogen in rice soils by algae as influenced by crop, CO_2, and inorganic substances, *Soil Sci.*, 70, 136—151, 1950.

341. **Rewari, R. B., Sundara Rao, W. V. B., and Venkataraman, G. S.**, Effect of nitrogen-fixing blue-green algae and bacteria on the yield of rice, *Indian J. Microbiol.*, 1, 82, 1961.

342. **Singh, R. N.**, *Role of Blue-Green Algae in Nitrogen Economy of India*, Indian Council Agric. Res., New Delhi, India, 1961.

343. **Aiyer, R. S., Venkataraman, G. S., and Sundra Rao, W. V. B.**, Effect of nitrogen-fixing blue-green algae and *Azotobacter chroococcum* on vitamin C content of tomato fruits, *Sci. Cult.*, 30, 557, 1964.

344. **Venkataraman, G. S., Dutta, N., and Natarajan, K. N.**, Studies on nitrogen fixation by blue green algae. I. Nitrogen fixation by *Cylindrospermum sphaerica* Prasad, *J. Indian Bot. Sci.*, 38, 114—119, 1959.

345. **Venkataraman, G. S.**, Algalization, *Phykos*, 5, 164—174, 1966.

346. **Okuda, A. and Yamaguchi, M.**, Nitrogen-fixing microorganisms in paddy soils. Characteristics of the nitrogen fixation in paddy soils, *Soil Plant Food*, 1, 102—104, 1955.

347. **Kobayashi, M., Takhashi, E., and Kawaguchi, K.**, Distribution of nitrogen-fixing microorganisms in paddy soils of southeast Asia, *Soil Sci.*, 104, 113—118, 1967.

348. **Watanabe, A.**, Studies on the blue-green algae as green manure in Japan, *Proc. Nat. Sci. India*, 35, 361—369, 1965.

349. **Watanabe, A. and Yamamoto, Y.**, Heterotrophic nitrogen-fixation by the blue-green alga *Anabaenopsis circularis.*, *Nature (London)*, 214, 5089, 1967.

350. **Watanabe, A., Lee, K. K., Alimagno, B. V., Sato, M., del Rosario, D. C., and deGusman, M. R.**, Biological nitrogen fixation in paddy field studied by in situ acetylene-reduction assays, IRRI Res. Paper, Series 3, Manila, Philippines, 1977.

351. **Yoshida, T., Roncal, R. A., and Bautista, E. M.**, Nitrogen fixation in a Philippine soil, Proc. 2nd Asian Soil Conf., Indonesia.

352. **El-Nawawy, N. A. S. and Hamdi, Y. A.**, Research on blue-green algae in Egypt, 1958—1972, in *Nitrogen Fixation by Free-Living Microorganisms*, Stewart, W. D. P., Ed., Cambridge University Press, New York, 1975.

353. **Materassi, R. and Balloni, W.**, Quelques observations sur la presence de microorganisms autotrophes fixateurs d'azote dans les rivieres, *Ann. Inst. Pasteur*, 3, 218—223, 1965.

354. **Bunt, J. S.**, A comparative account of the terrestrial diatoms of Macquarie Island, *Proc. Linn. Soc. N. S. W.*, 79, 40—53, 1954.

355. **Roger, P.**, Researches preliminaires sur des Cynophycees des sols de riziere du Senegal, Rapport de stage 1971—72, Orston, Dakar, Publ. Voneo, 1973.

356. **Raud, G.**, Study of nitrogen-fixing algae of the soil in the Lamto Savannah (Ivory Coast): elaboration of a new technique, *Ann. Univ. Abidjan Ser. E.*, 9, 305—336, 1978.

357. **Moore, A. W.**, Occurrence of non-symbiotic nitrogen-fixing organisms in Nigerian soils, *Plant Soil*, 19, 385—395, 1963.

358. **Calder, E. A.**, Nitrogen fixation in a Uganda swamp soil, *Nature (London)*, 184, 746, 1959.

359. **James, B. S., Hyun, S. K., Lee, E. W., and Watanabe, A.**, An observation on the effects of nitrogen fixing blue-green algae at acidic soil of rice fields in Korea, in *Taxonomy and Biology of Blue-Green Algae*, Desikachary, T. V., Ed., Symposium at Madras, India, 1970.

360. **Johnson, A.**, Blue-green algae of Malaysian rice fields, *J. Singapore Natl. Acad. Sci.*, 1, 30—36, 1969.

361. **Watanabe, A. and Yamamoto**, Algal nitrogen fixation in the tropics, *Plant Soil*, Special vol., 403—413, 1971.

362. **Shtina, E. A.**, Some regulation in the distribution of blue-green algae in soils, *Biol. Sinezeleni Khvodoroslei*, 2, 21—45, 1969.

363. **Gollerbakh, M. M. and Shtina, E. A.**, Soil algae, *Acad. Sci. USSR*, Moscow, 1969,

364. **Granhall, U.**, Nitrogen fixation by blue green algae in temperate soil, in *Nitrogen Fixation by Free Living Microorganisms*, Stewart, W. D. P., Ed., Cambridge University Press, New York, 1975.

365. **Beadle, N. C. W. and Tchan, Y. T.**, Nitrogen economy in semi-arid plant communities. I. The environment and general considerations, *Proc. Linn. Soc. N.S.W.*, 80, 62—70, 1955.

366. **Shields, L. M. Mitchell, C., and Drouet, F.,** Alga and lichen stabilized surface crust as a soil nitrogen source, *Am. J. Bot.,* 44, 489—498, 1957.

367. **Cameron, R. E.,** Fixation of Nitrogen by Algae and Associated Organisms in Semi-Arid Soils: Identification and Characterization of Soil Organisms, M.S. thesis, University of Arizona, Tucson, 1958.

368. **Cameron, R. E. and Fuller, W. H.,** Nitrogen fixation by some algae in Arizona soils, *Soil Sci. Soc. Am. Proc.,* 24, 353—356, 1960.

369. **Fuller, W. H., Cameron, R. E., and Raica, N.,** Fixation of nitrogen in desert soils by algae, Trans. 7th Cong. Intl. Soil Sci., Madison, Wis., 617—624, 1960.

370. **MacGregor, A. N. and Johnson, D. E.,** Capacity of desert algal crusts to fix atmospheric nitrogen, *Soil Sci. Soc. Am. Proc.,* 35, 843—844, 1961.

371. **Mayland, H. F. and McIntosh, T. M.,** Availability of biologically-fixed atmospheric nitrogen-15 to higher plants, *Nature (London),* 209, 421—422, 1966.

372. **Mayland, H. F. and McIntosh, T. N.,** Distribution of nitrogen fixed in desert algal-crust, *Soil Sci. Soc. Am. Proc.,* 30, 606—609, 1966.

373. **Mayland, H. F., McIntosh, T. N., and Fuller, W. H.,** Fixation of isotopic nitrogen on a semi-arid soil by algal crust organisms, *Soil Sci. Soc. Am. Proc.,* 30, 56—60, 1966.

374. **Snyder, J. M. and Wullstein, L. H.,** The role of desert cryptograms in nitrogen fixation, *Am. Midl. Nat.,* 90, 257—265, 1973.

375. **Holm-Hansen, O.,** Algae: nitrogen fixation by Antarctic species, *Science,* 139, 1050—1060, 1963.

376. **Alexander, V. A. and Schell, D. M.,** Seasonal and spatial variation of nitrogen fixation in the Banow Alaska tundra, *Arct. Alp. Res.,* 5, 77—88, 1973.

377. **Englund, B.,** Algal nitrogen fixation on the lava field of Heimae, Ireland, *Oecologia (Berlin),* 34, 45—56, 1978.

378. **Tichomirov, B. A.,** Dynamics of vegetation on arctic sport tundra, *Bot. Zurn.,* 42, 11—22, 1957.

379. **Stutz, C.,** Nitrogen Fixation in a High Arctic Ecosystem, Ph.D. dissertation, University of Alberta, Edmonton, 1973.

380. **Holm-Hansen, O.,** Isolation and culture of terrestrial and freshwater algae of Antarctica, *Phycologia,* 4, 43—51, 1964.

381. **Cameron, R. E.,** Soil microbial ecology of Valley of 10,000 Smokes Alaska, *J. Ariz. Acad. Sci.,* 6, 11—40, 1970.

382. **Shtina, E. A. and Bolyshev, N. N.,** Algae of the solontsy, *Bot. Zurn.,* 45, 11, 1960.

383. **Shtina, E. A.,** Some peculiarities of the distribution of nitrogen fixing blue-green algae in soils, in *The Taxonomy and Biology of Blue-Green Algae,* Desikachary, T. V., Ed., University of Madras, India, 1972.

384. **Jurgensen, M. F. and Davey, C. B.,** Nitrogen-fixing blue-green algae in acid forest and nursery soils, *Can. J. Microbiol.,* 14, 1179—1183, 1968.

385. **Metting, B.,** A Comparative Study of Algal Communities on Cultivated and Uncultivated Portions of a Schumacher Silt Loam, Ph.D. dissertation, Washington State University, Pullman, 1979.

386. **Zimmerman, W. L., Metting, B., and Rayburn, W. R.,** The occurrence of blue-green algae in Whitman Country, Washington soils, *Soil Sci.,* 130, 11—18, 1980.

387. **Marathe, K. V. and Anantani, Y. S.,** Observations on the algae of some Indian arid soils, *Botanique (Nagpur),* 3, 13—20, 1972.

388. **Marathe, K. V. and Anatani, Y. S.,** An example of algae as pioneers in the lithosphere and their role in rock corrosion, *J. Ecol.,* 63, 65—70, 1975.

389. **Reddy, G. N. and Giddens, J.,** Nitrogen fixation by algae in fescugran soil crusts, *Soil Sci. Soc. Am. Proc.,* 39, 654—656, 1975.

390. **Granhall, U. and Slender, H.,** Nitrogen fixation in a subarctic mine, *Oikos,* 24, 8—15, 1973.

391. **Lockyer, D. R. and Cowling, D. W.,** Non-symbiotic nitrogen fixation in some soil of England and Wales, *J. Br. Grassl. Soc.,* 32, 7—11, 1977.

392. **Witty, J. F. and Dart, J. M.,** The nitrogen economy of the Broadbalk experiments II. Biological nitrogen fixation, *Rothamsted Exp. Stn. Rep.,* 111—118, 1977.

393. **Torsvik, V. L.,** Norwegian IBP tundra biome annual report for 1971, 1971.

394. **Morne, A. J.,** The ecology of nitrogen fixation in Signy Island, South Orkney Islands, *Br. Antarct. Surv. Bull.,* 27, 1—18, 1972.

395. **Henriksson, E., Henriksson, L. E. and DaSilva, E. J.,** A comparison of nitrogen fixation by algae of temperate and tropical soils, in *Nitrogen Fixation by Free-Living Microorganisms,* Stewart, W. D. P., Ed., Cambridge University Press, New York, 1975, 199—206.

396. **Jahnke, E.,** Die Rolle Stickstoffbindender Blaualgen in mecklenbur gischen Boden, *Zentralbl. Bakteriol. Parasitenkd. Infektionskr. Hyg. Abt. 2,* 121, 636—642, 1967.

397. **Paul, E. A., Myres, R. J. K., and Rice, W. A.,** Nitrogen fixation in grassland and associated cultivated ecosystems, *Plant Soil,* Special vol., 495—507, 1971.

398. **Woodmansee, R. G.,** Additions and losses of nitrogen in grassland ecosystems, *BioScience,* 28, 448—453, 1978.

399. **Zimmerman, W. L.,** Selective Studies of Asymbiotic Nitrogen-Fixing Soil Cyanophytes in a Temperate Climate, M.S. thesis, Washington State University, Pullman, 1979.

400. **Fogg, G. E.,** Nitrogen fixation by blue-green algae, *Endeavour,* 6, 172—175, 1947.

401. **Fogg, G. E.,** Growth and heterocyst production in *Anabaena cylindrica* Lemm. II. In relation to carbon and nitrogen metabolism, *Ann. Bot.,* 13, 241—259, 1949.

402. **Fogg, G. E. and Stewart, W. D. P.,** In situ determinations of biological nitrogen fixation in Antarctica, *Br. Antarct. Sur. Bull,* 15, 29—46, 1968.

403. **Allen, M. B.,** Photosynthetic nitrogen fixation by blue-green algae, *Sci. Monthly,* 83, 100—106, 1956.

404. **Bunt, J. S.,** Nitrogen fixing blue-green algae in Australian rice soils, *Nature (London),* 192, 479—480, 1961.

405. **Venkataraman, G. S.,** Studies on nitrogen fixation by blue-green algae. II. Nitrogen fixation by *Cylindrospermum sphaerica* Prasad under various conditions, *Proc. Natl. Acad. Sci. India,* 31, 100—104, 1961.

406. **Varma, A. K., Mitra, A. K., and Venkataraman, G. S.,** Role of *Aulosira fertilissima* in the nitrogen economy of the soil, *Phykos,* 3, 29—32, 1964.

407. **Ohmori, M. and Hattori, A.,** Effect of ammonia on nitrogen fixation by the blue-green algae *Anabeana cylindrica, Plant Cell Physiol.,* 15, 131—142, 1974.

408. **Mosier, A. R.,** Inhibition of photosynthesis and nitrogen fixation in algae by volatile nitrogen bases, *J. Environ. Qual.,* 7, 237—240, 1978.

409. **Kennedy, I. R.,** Kinetics of acetylene and cyanide reduction by the nitrogen-fixing system of *Rhizobium lupini, Biochem. Biphys. Acta,* 222, 135—144, 1970.

410. **Granhall, O.,** Acetylene reduction by blue-green algae isolated from Swedish soils, *Oikos,,* 21, 330—332, 1970.

411. **Rinaudo, G., Balandreau, J., and Dommerques, Y.,** Algal and bacterial non-symbiotic nitrogen fixation in paddy soils, *Plant Soil,* Special vol., 471—480, 1971.

412. **Stanier, R. Y.,** The relationships between nitrogen fixation and photosynthetis, *Austr. J. Exp. Biol. Med. Sci.,* 52, 3—20, 1974.

413. **Balandreu, J. P., Miller, C. R., and Dommergues, Y. R.,** Diurnal variations of nitrogenase activity in the field, *Appl. Microbiol.,* 27, 662—665, 1974.

414. **Allison, F. E., Hoover, S. R., and Morris, H. J.,** Physiological studies with the nitrogen-fixing algae, *Nostoc muscorum, Bot. Gaz.,* 98, 433—463, 1937.

415. **Fay, P.,** Factors influencing dark nitrogen fixation in a blue-green alga, *Appl. Environ Microbiol.,* 31, 376—379, 1976.

416. **Alimagno, B. V. and Yoshida, T.,** In situ acetylene-ethylene assays of biological nitrogen fixation in lowland rice soils, *Plant Soil,* 47, 239—244, 1977.

417. **Renaut, J., Sasson, A., Pearson, H. W., and Stewart, W. D. P.,** Nitrogen-fixing algae in Morocco, in *Nitrogen Fixation by Free-Living Microorganisms,* Stewart, W. D. P., Ed., Cambridge University Press, New York, 1975.

418. **Taha, E. E. M. and El-Refai, A. M. H.,** Physiological and biochemical studies on nitrogen-fixing blue-green algae I. On the mat of cellular and extracellular nitrogenous substances formed by *Nostoc commune, Arch. Mikrobiol.,* 41, 307—312, 1962.

419. **Taha, E. E. M. and El-Refai, A. M. H.,** Physiological and biochemical studies on nitrogen-fixing blue-green algae. II. The role of calcium, strontium, cobalt, and molybdenum in the nitrogen fixation of *Nostoc commune, Arch. Mikrobiol.,* 43, 67—75, 1962.

420. **Rychert, R. C. and Skujins, J.,** Nitrogen fixation by blue-green algae-lichen crusts in the Great Basin Desert, *Soil Sci. Am. Proc.,* 38, 768—771, 1974.

421. **Pattnaik, H.,** Growth and nitrogen fixation by *Westiellopsis prolifica* Jent, *Ann. Bot.,* 30, 231—238, 1966.

422. **Jones, K.,** The effects of moisture on acetylene reduction by mats of blue-green algae in subtropical grasslands, *Ann. Bot.,* 41, 801—806, 1977.

423. **Kallio, P. and Kallio, S.,** Nitrogen fixation by free-living microorganisms in the Kevo district, *Rep. Kevo Subarctic Res. Stat.,* 12, 28—33, 1975.

424. **Granhall, U. and Henriksson, E.,** Nitrogen-fixing blue-green algae in Swedish soils, *Oikos,* 20, 175—178, 1969.

425. **Rodgers, G. A.,** Nitrogenase activity in *Nostoc muscorum,* recovery from desiccation, *Plant Soil,* 46, 671—674, 1977.

426. **Singh, R. N.,** The fixation of elementary nitrogen by some of the commonest blue-green algae from paddy soils of the United Provinces and Bihar, *Indian J. Agric. Sci.,* 12, 743—746, 1942.

427. **Tret'Yakova, A. N.,** Comparative study of nitrogen-fixing blue-green algae taken from various soils of the USSR, *Mikrobiologiya,* 34, 3, 1965.

428. **Brannon, M. S.,** Factors affecting growth and distribution of Myxophyceae in Florida, *Proc. Fla. Acad. Sci.,* 8, 296—303, 1945.

429. **Lund, J. W. G.,** The marginal algae of certain ponds, with special reference to the bottom deposits, *J. Ecol.,* 30, 245—283, 1942.

430. **Shields, L. M. and Durrell, L. W.,** Algae in relation to soil fertility, *Bot. Rev.,* 30, 92—128. 1964.
431. **Henriksson, E., Enckell, P. H., and Henriksson, L. E.,** Determination of the nitrogen fixing capacity of algae in soil, *Oikos,* 23, 420—423, 1972.
432. **Singh, S. P.,** Succession of blue-green algae on certain sites near Varanasi, *Indian J. Microbiol.,* 18, 128—130, 1978.
433. **Smith, R. V. and Evans, M. C. W.,** Soluble nitrogenase from vegetative cells of the blue-green alga *Anabaena cylindrica, Nature (London)* 225, 1253—1254, 1970.
434. **Haystead, A. and Stewart, W. D. P.,** Characteristics of the nitrogenase system of the blue-green algae *Anabaena cylindrica, Arch. Mikrobiol.,* 82, 325—326, 1972.
435. **Gallon, J. R., LaRue, T. A., and Kuz, W. G. W.,** Characteristics of nitrogenase activity in broken cell preparations of the blue-green algae *Gloeocapsa* sp. LB 795, *Can. J. Microbiol.,* 18, 327—332, 1972.
436. **Dhar, N. R. and Bhat, G. N.,** Influence of light sensitivity, organic matter and phosphate on: A) nitrogen fixation and B) availability of P_2NO_5 in the presence and absence of *Anabaena naviculoides* and *Chlorella pyrenoidosa, Proc. Natl. Acad. Sci. India,* Sect. A, 35, 309—326, 1965.
437. **Allen, M. B. and Arnon, D. I.,** Studies on nitrogen fixing blue-green algae II. The Na requirements of *Anabaena cylindrica* Lemm, *Physiol. Plant,* 8, 653—660, 1955.
438. **Allen, M. B. and Arnon, D. I.,** Studies on nitrogen-fixing blue - green algae. I. Growth and nitrogen fixation by *Anabaena cylindrica, Lemm. Plant Physiol.,* 30, 336—372, 1955.
439. **Eyster, C.,** The micro-element nutrition of *Nostoc Muscorum, Ohio J. Sci.,* 58, 25—33, 1958.
440. **Wolfe, M.,** The effect of molybdenum upon the nitrogen metabolism of *Anabaena cylindrica.* I. A study of the molybdenum requirement for nitrogen fixation and for nitrate and ammonia assimilation, *Ann. Bot.,* 18, 301—308, 1954.
441. **Venkataraman, G. S.,** The effect of molybdenum on the rate of growth and nitrate absorption by *Clylindrospernum sphaerica, Curr. Sci.,* 27, 306—307, 1958.
442. **Jacobs, R. and Lind, O.,** The combined relationship of temperature and molybdenum concentration to nitrogen fixation by *Anabaena cylindrica, Microbiol., Ecol.,* 3, 205—217, 1977.
443. **Subramanyan, R.,** Some observations on the utilization of blue-green algae mixtures in rice cultivation in India, in *The Taxonomy and Biology of Blue-Green Algae,* Desikachary, T. V., Ed., University of Madras, India, 1972.
444. **Ley, S. H.,** The effect of nitrogen-fixing blue-green algae on the yields of rice plant, *Acta Hydrobiol.,* 4, 440—444, 1959.
445. **Lay, S. H., T sing-Chuan, Y., Fu-jui, L., Lith-Mei, W., and Shi - Kiung, T.,** The nitrogen fixation of some blue-green algae from Chinese rice fields, *Acta Hydrobiol.,* 4, 445—450, 1959.
446. **Ley, S. H., Tsing-Chung, Y., Fu-Jui, L., Lith-Mei, W., and Shi - Kiung, T.,** The effect of nitrogen-fixing blue-green algae on the yields of rice plant, *Acta Hydrobiol.,* 4, 451—453, 1959.
447. **Mishustin, E. N.,** Biological nitrogen fixation in agriculture and prospectus of utilizing nitrogen-fixing blue-green algae in agriculture, *Proc. Sci. Adv. Comm. Physiol. Biochem. Mikroorganisms. Acad. Sci. USSR,* 1—47, 1964.
448. **Watanabe, A., Hattori, A., Fiyita, T., and Kiyohara, T.,** Large scale culture of a blue-green alga *Tolypothrix tenuis* utilizing hot spring and natural gas as heat and carbon dioxide source, *J. Gen. Appl. Microbiol.,* 5, 51—57, 1959.
449. **Watanabe, A.,** Effect of nitrogen-fixing blue-green algae *Tolypothrix tenuis* on the nitrogenous fertility of paddy soil and on the crop yield of rice plant, *J. Gen. Appl. Microbiol.,* 8, 85—91, 1962.
450. **Huang, Chi-Yng,** Effects of nitrogen-fixing activity of blue-green algae on the yield of rice plants, *Bot. Bull. Acad. Sci.,* 19, 41—52, 1978.
451. **Naway, A. S., Lotei, M., and Fahmy, M.,** Studies on the ability of some blue-green algae to fix atmospheric nitrogen and their effect on growth and yield of paddy soils, *Agric. Res. Rev.,* 36, 308—363, 1958.
452. **Perminova, G. N.,** Effect of blue-green algae on the development of microorganisms in the soil, *Mikrobiologya,* 33, 472—476, 1964.
453. **Shtina, E. A.,** Fixation of free nitrogen in blue-green algae, in *The Ecology and Physiology of Blue-Green Algae,* Federov, V. D. and Tellichenko, M. M., Moscow University Press, Moscow, 1965.
454. **Relwani, L. L. and Subramanyan, R.,** Role of blue-green algae, chemical nutrients and partial soil sterilization on paddy yield, *Curr. Sci.,* 32, 441—443, 1963.
455. **Subramanyan, R. and Sahay, M. N.,** Observation of nitrogen fixation by some blue-green algae and remarks on its potentialities in rice culture, *Proc. Indian Acad.Sci.,* 60, 145—154, 1964.
456. **Subramanyan, R. and Sahay, M. N.,** Observation on nitrogen fixation and organic matter produced by *Anabaena circinalis* and their significance in rice culture, *Proc. Indian Acad. Sci.,* 61, 164—169, 1965.
457. **Subramanyan, R., Relwani, L. L., and Manna, G. B.,** Observations on the role of blue-green algae on rice yield compared with the conventional fertilizers, *Curr. Sci.,* 33, 485—486, 1964.
458. **Subramanyan, R., Relwani, L. L., and Manna, G. B.,** Role of blue-green algae and different methods of partial soil sterilization on rice yield, *Proc. Indian Acad. Sci.,* 60B, 293—297, 1964.

459. **Subramanyan, R., Relwani, L. L., and Manna, G. B.,** Fertility buildup of rice field soils by blue-green algae, *Proc. Indian Acad. Sci.*, 62, 252—272, 1965.

460. **Subramanyan, R., Relwani, L. L., and Manna, G. B.,** Nitrogen enrichment of rice soils blue-green algae and its effects on the yield of paddy, *Proc. Natl. Acad. Sci.*, 35, 283—286, 1965.

461. **Jagnow, G.,** The influence of irrigation and crop rotation on the humus and nitrogen balance and soil productivity in the Sodan Gezira, *Z. Pflanzen, Boden Kd.*, 134, 20—32, 1973.

462. **Relwani, L. L. and Manna, G. B.,** Effect of blue-green algae in combination with urea on paddy yield, *Curr. Sci.*, 33, 687, 1964.

463. **Jha, K., Ali, M. A., Singh, R.N., and Bhattacharya, R.,** Increasing rice production through the inoculation of *Tolypothrix tenuis* a nitrogen-fixing blue-green alga, *J. Indian Soc. Soil Sci.*, 13, 161—166, 1965.

464. **Watanabe, A.,** On the mass-culturing of a nitrogen-fixing blue-green alga, *Tolypothrix tenuis, J. Gen. Appl. Microbiol.*, 5, 85—91, 1959.

465. **Watanabe, A.,** Distribution of nitrogen-fixing blue-green algae in various areas of south and east Asia, *J. Gen. Appl. Microbiol.*, 5, 21—29, 1959.

466. **Fay, P., Stewart, W. D. P., Walsby, A. E., and Fogg, G. E.,** Is the Heterocyst the site of nitrogen fixation in blue-green algae?, *Nature (London)*, 220, 810—821, 1968.

467. **Fogg, G. E.,** Blue green algae, in *Handbook of Microbiology*, Laskin, A. I. and Lechevalier, H. A., Eds., Vol I, 2nd ed., CRC Press, Boca Raton, Fla., 1978, 347—367.

468. **Wolk, C. P., Austin, S. M., Bortines, J., and Galonsky, A.,** Autoradiographic localization of ^{13}N after fixation of ^{13}N-labelled nitrogen gas by a heterocyst forming blue green alga, *J. Cell. Biol.*, 61, 440—453, 1974.

469. **Van Gorkon, H. J. and Donze, M.,** Localization of nitrogen fixation in *Anabaena, Nature (London)*, 234, 231—232, 1971.

470. **Stewart, W. D. P. and Lex, M.,** Nitrogenase activity in blue green algae *Plectonema boryanum* strain 594, *Arch. Mikorbiol.*, 73, 250—260, 1970.

471. **Stewart, W. D. P.,** Blue green algae, in *A Treatise on Dinitrogen Fixation*, Hardy, R. W. E. and Silver, W. S., Eds., John Wiley & Sons, New York, 1977, 63—123.

472. **Wyatt, J. J. and Silvey, J. K. G.,** Nitrogen fixation by *Gloeocapsa, Science*, 165, 908—909, 1969.

473. **Holst, R. W. and Yopp, J. H.,** Studies of the *Azolla-Anabaena* symbiosis using *Azolla mexicana*. I. Growth in nature and laboratory, *Am. J.*, 69, 17—25, 1979.

474. **Holst, R. W. and Yopp, J. H.,** Environmental regulation of nitrogenase and nitrate reductase as systems of nitrogen assimilation in the *Azolla mexicana* and *Anabaena azollae* symbiosis, *Aquat. Bot.*, 7, 369—384, 1979.

475. **Peters, G. A.,** Studies on *Azolla-Anabaena azollae* symbiosis, *Proc. Int. Sym. Nitrogen Fixation*, Washington State University Press, Pullman, 1975, 592—610.

476. **Peters, G. A. and Mayne, B. C.,** The *Azolla-Anabaena azollae* relationship II. Localization of nitrogenase activity as assayed by acetylene reduction, *Plant Physiol.*, 53, 820—824, 1974.

477. **Peters, G. A., Evan, W. R., and Toi, R. E., Jr.,** *Azolla-Anabaena azollae* relationship. IV. Photosynthetically driven, nitrogenase-catalyzed H_2 production, *Plant Physiol.*, 51, 332—336, 1976.

478. **Sculthorpe, C. D.,** *The Biology of Aquatic Vascular Plants*, Edward & Arnol, London, 1967.

Chapter 2

EXPERIMENTAL, METHODOLOGICAL, AND ANALYTICAL APPROACH TO STUDY THE EFFECTS OF PESTICIDES ON NITROGEN CYCLE

Rup Lal

TABLE OF CONTENTS

I. INTRODUCTION

Pesticides are widely used in agriculture to reduce the losses in food production caused by weeds, insects, and fungal or bacterial diseases. The importance of these pesticides as pest controllers is unlikely to diminish in the foreseeable future, despite great advances in biological pest control. Over the next 10 years food production has to be increased by 100% to meet worldwide needs. Thus, there will be a continuing demand for new chemicals — more specific, suitable for integrated pest control, less vulnerable to resistance, and of low mammalian toxicity.

It cannot be assumed that a potential hazard does not exist when large quantities of "antibiological" chemicals are introduced into the environment; new compounds developed as pesticides should be extensively screened for ecological side effect. Any such screening program must include ecological studies on the soil environment since, even if not directly applied to it, pesticides will eventually arrive in the soil one way or another. Since microorganisms are responsible for processes such as organic matter turnover, nitrogen transformation, and nitrogen fixation, (which have a major influence on plant growth) the benefit of pesticides as pest combatants could be outweighed by detrimental effect on soil microorganisms, and hence soil fertility or crop production.

Several papers have appeared on the effects of pesticides on microorganisms in soil.[1-17] It is evident from these papers that very little attention has been given to study the effects of pesticides on nitrogen transformations. This is mainly due to the lack of standard methodology which has not emerged even after decades of research in this field. In this chapter the experimental, methodological, and analytical procedures to study the effects of pesticides on microorganisms involved in nitrogen cycle have been outlined. Some of these techniques described have not been specifically designed to study the effects of pesticides on these microbes but they can be manipulated and exploited to a maximum extent to understand the pesticide interactions with such microbes. If the ecology of a particular organism is known and a method is available to enumerate its population and activities, one only has to know about the pesticide used, conditions where such organisms grow, along with the application rates and residue levels. A careful treatment and subsequent analysis can give the data regarding the impact of pesticides on a particular transformation of nitrogen cycle.

II. GENERAL CONSIDERATIONS

A. Soil Treatment With Pesticides

In routine experiments to determine the effects of pesticides on soil microorganisms it is desirable that the soil be uniformly treated with pesticides. To achieve this, soils are spread as a 2.5 cm layer on trays covered with polyethylene sheets and sprayed with the laboratory pot sprayer.[18] After spraying, the soils are mixed thoroughly. The pesticides should always be sprayed into moist soil. The total volume of water and pesticide used is calculated to produce the desired final soil moisture content, usually 60% moisture holding capacity. Granular formulation quantities of pesticides are ground in a pestle and mortar with a small amount of dry soil. The soil pesticide mixture is then spread as evenly as possible over the soil to be treated and thoroughly mixed.

Pesticide concentration used will depend on the purpose of the experiment. For routine work, however, soils are sprayed at the manufacturer's highest recommended field rate and at ten times that rate.

B. Dilution Plate Technique

Samples of control or treated soil are placed inside Cryovac bags with 100 mℓ sterile water. The soil suspension is then treated in a stomacher to remove and suspend microorganisms attached to the root surface. It is advisable to use the double Cryovac to avoid puncture during treatment in the stomacher. This initial suspension of soil is diluted tenfold by pipetting 10 mℓ into 90 mℓ of sterile water and shaking the suspension by hand for 1 min. A 1 mℓ aliquot is transferred to 9 mℓ sterile water in a test tube wih a polyethylene cap and the suspension mixed on the vortex shaker. This suspension is further diluted in the same way and the procedure repeated to produce the required range of dilution. Greaves et al.[18] used a dilution of 10^{-4}, 10^{-5}, and 10^{-6} for bacteria. Suitable volumes of the diluted solution are pipetted onto the surface of plates of a suitable agar medium. Before use the plates are kept at 37°C for 2 to 3 days to dry any surface moisture on the agar and promote rapid absorption of the soil dilution into agar. Three replicate plates are prepared for each dilution. Plates are incubated at 19 to 20°C for 14 days. After incubation the numbers of bacteria are counted — spore-forming bacteria recounted in soil dilution which have been heated at 80°C for 10 min in a water bath.

One of the many disadvantages of the plate-dilution method for counting microbial propagules is the large imput of manpower and materials. This stage can be overcome by the use of a miniaturized most-probable number technique (MPN), which also increases the processing of samples and allows their handling at one time. In this method about 100 $\mu\ell$ of the medium is dispensed into all the wells on a microtitration plate, with the exception of those in the first. Each well of the first row receives about 100 $\mu\ell$ of soil suspension from Pasteur pipettes. This suspension is prepared by treating 1 g soil in 100 mℓ sterile deionized water for 1 min in the stomacher and then 10 mℓ of this suspension is diluted with 90 mℓ of sterile deionized water.

Preparation of a dilution series from the soil suspension in the first row of wells is achieved using 25 $\mu\ell$ microdilutors. The heads of these microdilutors are calibrated to pick up the exact volumes of solution by capillary action. There is some risk when using soil suspension, especially in organic soils where particles may lodge in the microdilutor head. The microdilutor rods are dipped into the suspension in the first row of wells, rotated 15 times in approximately 6 sec to mix the suspension without causing frothing and then transferred to the medium in the second row of the wells. Microdiluters are rotated again and transferred to the third row. This is repeated to affect transfers to all 12 rows of wells. Each transfer of 25 $\mu\ell$ from one well to the next produces a fivefold dilution of the suspension carried over. When the dilution series down all 12 rows of wells in a microtitration plate have been achieved, the lid of the plates are replaced and the plates are incubated in dark at 19 $\pm$ 2°C for 5 days.

After incubation, any bacterial growth in the wells, shown by the turbidity of the medium or aggregated growth at the bottom of well, is recorded. This is facilitated by using an illuminated petri-dish viewer. Where there is doubt about the growth in any well, a sample is taken and examined using a microscope. When it is necessary to incubate for longer periods than 5 days or when higher incubation temperatures are required it is advisable to place the plates inside plastic boxes with a damp filter paper in the base. The lids of the plates are secured in places with rubber bands. These precautions reduce moisture loss from the medium in the wells and prevent drying out.

C. Evaluation of Data on Growth

An estimation of the number of bacteria in the soil suspension is made by recording the presence or absence of growth in all the replicates. The major advantage of plate dilution and MPN methods is to use them to count different nonsymbiotic nitrogen-fixers by using a selective medium. In a systematic analysis of the data it becomes evident that most of the described effects cannot be generalized. This is because in the natural environment there exists no organisms that live in isolation. Between the terrestrial ecosystem a network of interactions keeps the population in a state of dynamic equilibrium. Thus, any effect on microbes has to be seen in the light of these environments. Domsch et al.[19] in an attempt to generalize data proposed different types of patterns of the effects of pesticides on microbes. Figure 1 illustrates two principal types of reversible inhibition. In one case the base line of the given population of function (untreated control sample) remains constant (in an ideal case) or a depression is induced (treated sample). In the second case a response is induced at the beginning of the monitoring period (e.g., nitrate production in a nitrification experiment) and it is irrelevant whether the induction is triggered at the time of or after the application of the pesticide. In both types the inhibition can be described by the maximum depression (the greatest observed difference between treated and untreated samples), and the maximum delay (equals the greatest observed time-lag between treated and untreated samples). In natural environments temperature has to be taken into account as a driving variable. Below the respective optimum temperature a Q_{10} can be taken as a reasonable approximation. Consequently, between a 50 to 25°C range of delay period can be calculated. The "weighted" delayed periods can thus be adapted to realistic temperatures. Figure 2 shows the affects which temperature might have on the three delay ranges, negligible, tolerable, and critical. Figure 3 shows two principal types of persistent inhibition. In both types the situation at the end of the monitoring period can be described as a deficit which is equal to the residual difference between treated and untreated samples. From the principal ability and restitution it can be concluded that, as a rule, depressions are reversible and the monitoring period is a most important criterion for the assessment of toxic effects. For instance, a period of 30 days is minimum for the recognition of persistent inhibition as the system will return to normal with a monitoring period of 30 days only. This further means that less than 30 days is certainly inadequate for the recognition of a persistent effect.

III. AMMONIFICATION

Ammonification, as a part of nitrogen cycle is a useful criterion to study potential side

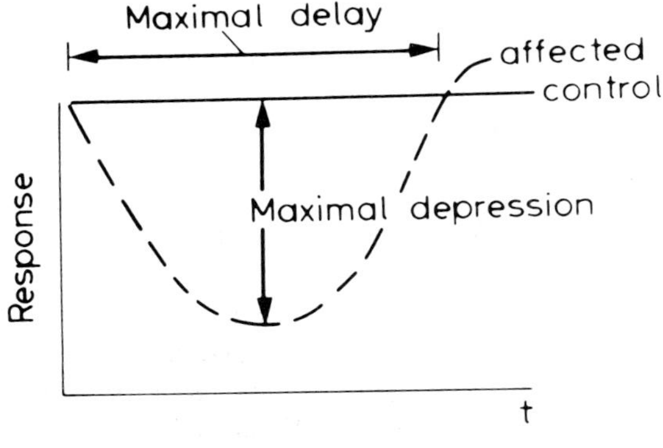

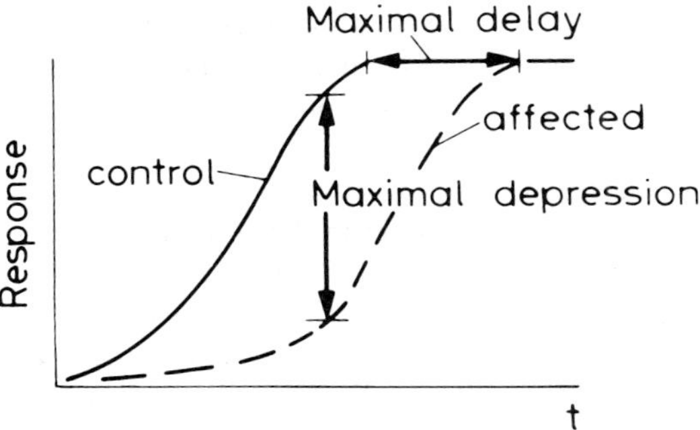

FIGURE 1. Principal types of reversible responses characterized by depression of microbial test parameters and delay
periods of recovery. (From Domsch, K. H., Jagnow, G., and
Anderson, T. H., *Residue Rev.*, 86, 65—105, 1983. With
permission.

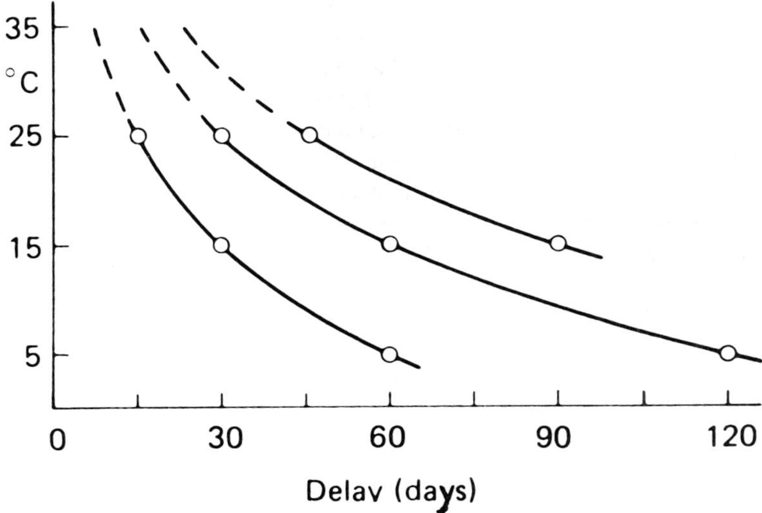

FIGURE 2. Relationship between recovery and delay periods, respectively and
ambient temperature in natural environments, assumed doubling time 10 days, assumed Q_{10} = 2. From left to right ranges of critical, tolerable and critical delay
periods. (From Domsch, K. H., Jagnow, G., and Anderson, T. H., *Residue Rev.*,
86, 65—105, 1983. With permission.)

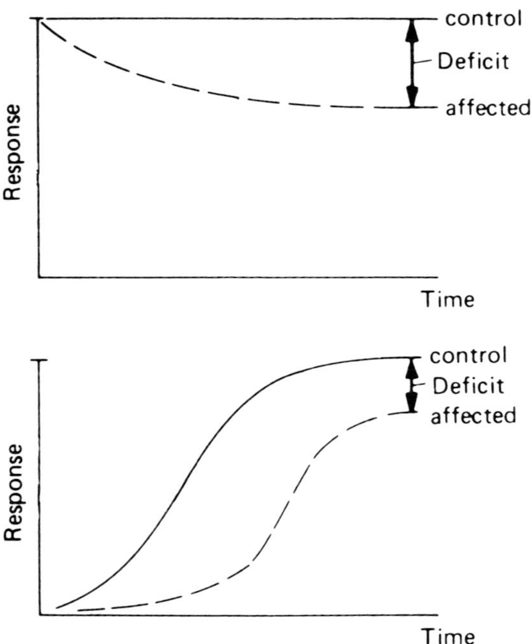

FIGURE 3.   Principal types of persistent responses characterized by deficits at the end of the monitoring period. (From Domsch, K. H., Jagnow, G., and Anderson, T. H., *Residue Rev.*, 86, 65—105, 1983. With permission.)

effect of pesticides. Ammonification is the indicator for the release of nitrogen bound to organic matter and its availability for plant nutrition. As mentioned earlier (see Chapter 1) proteolysis of the larger protein molecules occurs first before deamination leading to the formation of ammonia. Few techniques have been developed for evaluating the microflora responsible for proteolysis and ammonification. Pochon and Chalvignac[20] mixed particles of sterilized animal tissues with the soil, isolated the organisms developing on them,, and tested each for its ability to lyse proteins. Coagulated blood serum in petri dishes, inoculated with aliquots of a serial dilution of soil and isolation of the colonies which caused liquefaction, was the method used by Chalvignac.[21] Subsequently Lajudie and Chalvignac[22] used a selective gelatin culture medium and the most probable number method for determining numbers of protein liquefying microorganisms.

While the above methods enable both isolation and computation of the number of microorganisms, the best method for evaluating the activity of proteolytic organisms in pesticide-treated soil is by incubating the soil with uniformly ^{14}C-labeled protein and collecting the $^{14}CO_2$ evolved. The extent and the rate of breakdown of the protein will be reflected by CO_2 evolution irrespective of the number of steps and types of microorganisms involved.

Ammonification measurements in soil may easily be confounded by absorption of the ammonium ions into clay particles. Thus, Kauffmann and Chalvignac[23] mixed a variety of nitrogenous substrates into sand and inoculated the sand with microflora extracted from soil, to obviate the absorption of ammonium ions into clay.

Greenwood and Lees[24] modified the Lees-Quastel apparatus to measure the extent of oxygen utilization during studies on ammonification and nitrification of amino acids. Since absorption and nitrification can contribute significant errors in ammonification studies by such methods, liquid elective media containing an amino acid may be used in conjuction with the most probable number method to minimize error. The disappearance of the amino acid and appearance of ammonia in each of the dilutions of soil can be recorded as a function of time and dilution.

Greaves et al.[18] have described a method for the analysis of ammonium, nitrite, and nitrate. Samples (25 g) of treated and control soils are extracted by shaking with 50 mℓ of 2 MKC1 for 1 hr on the shaker. The extract is filtered through a Whatman No. 1 paper into 100 mℓ conical flasks and the filtrate analyzed on the autoanalyzer. Ammonium and nitrate are analyzed simultaneously at 20 samples per hour on both channels of the autoanalyzer. In such a case where nitrate is reduced to nitrite the color measured by the autoanalyzer is equivalent to $NO_2^- + NO_3^-$ in the sample and NO_3^- is found by subtraction.

The release of ammonium from organic materials can also be studied by the addition of 0.5% lucerne-meal to soil. If ammonification is not affected, the two steps of nitrification can be studied in the same experiment. If ammonification is inhibited, a separate experiment is required to study nitrification independently.[25] This is achieved by addition of ammonium sulfate (100 ppm N) to the soil and following the disappearance of NH_4^+ and the appearance of NO_2^- and NO_3^-. The study of ammonification needs to be carried out until an equilibrium is reached between ammonification and nitrification, but in any case for not less than 4 weeks. The study of the nitrification of ammonium sulfate has to be continued until either the added substrate has been covered or until equilibrium has been reached between ammonification from soil organic matter and nitrification.

IV. NITRIFICATION

The usual approach to the microbiology of nitrification is indirect. The accuracy and sensitivity of the MPN for quantification of these bacteria is unknown, as no standard exists. Qualitatively, the MPN is limited in reflecting the diversity of nitrifiers because of the elective nature of the method. It is also difficult to plate-out the nitrifiers on selective media. The plating techniques of Walker[26] have not come into widespread usage despite the presently widespread and elevated interest in both the enumeration and isolation of nitrifiers. Direct plating for isolation, even with strictly inorganic media, is virtually useless because the organic materials introduced with the inoculum permit the growth of non-nitrifying heterotrophs. Further the isolation of nitrifiers is a tedious process requiring careful and extensive serial enrichment of cultures which increases the chances of contamination.

Nitrifiers also grow slowly in the culture media which makes them susceptible to contamination. Further, this makes it difficult to carry out studies on the affects of pesticides for a longer period. Only a few strains of *Nitrosomonas* and *Nitrobacter* can be maintained without contamination in pure culture forms. To extend the methodology evolved with *Nitrosomonas* and *Nitrobacter* to other nitrifying bacteria, it is necessary to first understand the nature of such microbes. The nitrifiers found in soil sewage and aquatic environments should also be clearly understood before making any attempt to study the effects of pesticides on nitrification by such microorganisms.

The simplest and most convenient method for studying the effects of pesticides on nitrification is by incubation of the soil in containers with the addition of known amounts of ammonium ions. At set intervals of time replicate flasks of each treatment are removed and the soil analyzed for residual ammonium ions as well nitrite and nitrate ions. Lees and Quastel[27] described a system of perfusion of soil with an aerated liquid and since then many variations and modifications of the original Lees-Quastel apparatus have been described. The device consists of a column of soil connected to a reservoir containing a solution of ammonium ions (Figure 4). Droplets of the solution are induced to flow to the soil column under a slight positive air pressure. After percolating through the soil the liquid returns to the reservoir. A sampling port in the reservoir permits removal of aliquots for analysis of NH_4^+, NO_2^-, and NO_3^- ions.

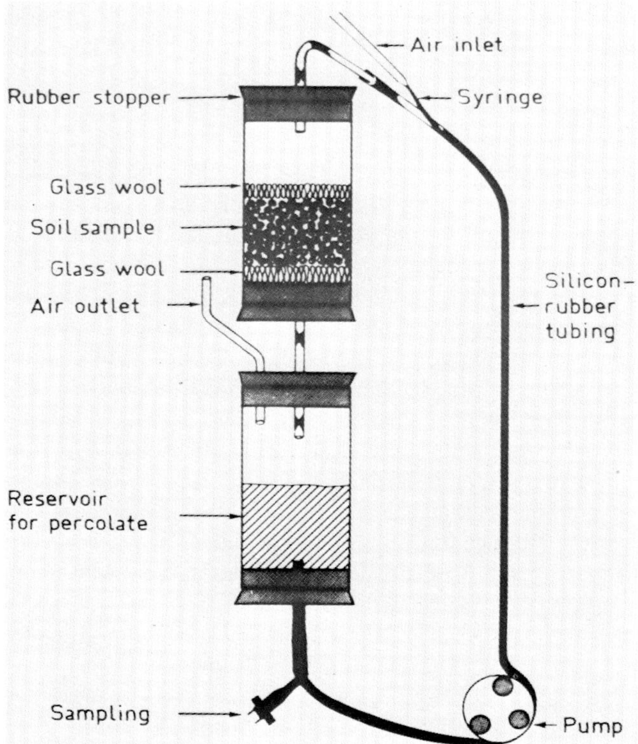

FIGURE 4.   Schematic representation of soil percolation apparatus used for study of nitrification. (From Torstensson, L., *Swedish J. Agric.*, 4, 151—160, 1974. With permission.)

Many advantages and disadvantages of the method were listed by Lees and Quastel.[27] Among the advantages is the fact that the soil is kept at a constant and evenly distributed water content; temperature variation in the soil are minimized; soil is undisturbed during the experiment and gassing (by air or any other atmosphere) is guaranteed; sampling and the addition of any inhibitor if required, is facilitated, and the soil and perfusate can be analyzed individually. The disadvantages of the system are: the removal of perfusate for analysis decreases the total volume of the perfusate without a compensating decrease in the amount of soil and also the soil is always in danger of becoming water logged. The latter disadvantage is a very serious hazard as any worker who has used the system knows. The apparatus also requires careful construction and maintenance otherwise liquid accumulates above the soil or does not percolate at all.

Estimating nitrification rates in samples from natural environments by measuring absolute changes in ammonium, nitrite, or nitrate is often very difficult. Other organisms can be simultaneously producing and utilizing these ions. The rates of change in the concentration of these ions can be small in relation to the absolute concentration; and the absolute concentrations of the compound may be near the detection limits of standard analytical methods. To overcome this sensitive problem it has been proposed by Billen[28] and Samville[29] that nitrification should be estimated indirectly by measuring bicarbonate uptake by nitrifying bacteria. The underlying assumption being that there is a constant stoichiometric ratio between the rate of substrate oxidation and rate of bicarbonate uptake for a given species of nitrifier. It must also be assumed in any application of these methods that the uptake ratio is independent of the growth rate and environmental conditions. Hall[30] questioned whether

Table 1
LIST OF SYMBOLS AND DEFINITION USED IN THE NITRIFICATION MODEL

Symbols (original)	Symbols (adpated)	Definitions
NH_4 free	S_0	Concentration of NH_4^+-ions in solution
NH_4 sorb	S_1	Concentration of NH_4^--ions
NH_4 sorb (0)	S_1 (0)	Initial concentration of NH_4^+-ions adsorbed to clay minerals
NH_4 sorb (end)	S_1 (end)	Concentration of residual non-oxidization NH_4^+-ions
NO_2	S_2	Concentration of NO_2^--ions
NO_3	S_3	Concentration of NO_3^--ions
E_1	E_1	Oxidation rate (linear proportional to biomass) of *Nitrosomonas* sp.
E_2	E_2	Oxidation rate (linear proportional biomass) of *Nitrobacter* sp.
pH (0)	pH(0)	Initial H^+-ion concentration
c	K_3	Rate of pH change
k free	K_1	NH_4^+-exchange parameter
k_1^E	K_2	Parameter expressing losses due to assimilation
$a_{1,2}$ (pH)	$u_{1,2}$ (pH)	Functions of pH-dependent optimal growth
$K_{1,2}$	$l_{1,2}$	Monod parameters
$b_{1,2}$	$d_{1,2}$	Parameters expressing losses due to natural death
	$H_{1,2}$	Functions of residual activities
	c_{ij}	Constants to shape the sigmoid inhibition due to action of different herbicides

From Domsch, K. H. and Paul, W., *Arch. Microbiol.*, 97, 283—301, 1974. With permission.

these ratios were constant. He observed that the biocarbonate uptake method consistently underestimated observed rates of nitrification. Belser,[31] while studying the validity of this method, pointed out that the method could not be used to measure the natural populations of nitrifiers since there was no stoichiometric relationship between biocarbonate uptake and substrate oxidation rate.

The chemoautotrophic nitrifiers are relatively sensitive to change in their environment. Because of this, the influence of pesticides on the process of nitrification has been frequently investigated on these microorganisms. Domsch and Paul[6] emphasized the need to consider the following points in order to evaluate the effects of pesticides on nitrification.

1. Oxidation of NH_4^+ ions converts nitrogen to a leachable form, an inhibition of the NH_4^+ oxidation is, therefore, from the standpoint of plant production and is not a negative effect.
2. Inhibition of NO_2^- oxidation can lead to an undesired accumulation of NO_2^- ions.
3. Inhibition effects which have been shown in vitro, have a soil ecological significance only when inhibition continues for an appreciable period of time.
4. Physical, chemical and biological properties of different soils influence the course of nitrification and thereby also influence the extent of inhibition.

From these statements it is clear that only the systematic analysis of parameters relevant to the nitrification process can lead to meaningful conclusion concerning the pesticide effects. To approach this goal, a methematical model for the nitrification process was developed by Paul and Domsch.[32] Following are the equations which were described in this model of nitrification. These equations include functions of H_1 and H_2 which define residual biological activities. Table 1 contains the symbols and definitions as they were used in the original paper.

$$S_0 = -K_1 S_0 [S_1(0) - S_1]$$

$$S_1 = K_1 S_0 [S_1(O) - S_1] - E_1 H_1 [S_1 - S_1 \text{ (end)}]$$

$$S_2 = K_2 E_1 H_1 [S_1 - S_1 \text{ (end)}] - E_2 H_2 S_2$$

$$S_3 = E_2 H_2 S_2$$

$$pH(t) = pH(O) - K_3 t$$

$$E_1 = \left\{ \frac{u_1(pH) [S_1 - S_1 \text{ (end)}]}{l_1 + S_1 - S_1 \text{ (end)}} - d_1 \right\} E_1 H_1$$

$$E_2 = \left\{ \frac{u_2(pH) S_2}{l_2 + S_2} - d_2 \right\} E_2 H_2$$

E_1 and E_2 are to be interpreted as enzyme activities, which are very likely proportional to the biomass m_1 and m_2.[33] Since in their experiment no attempts have been made to determine m_1 and m_2, a formal oxidation rate with the dimension one per unit has been chosen as state variable rather than biological variable biomass.

For $H_1 = H_2 = 1$, above equations are the basic model with no inhibition. The activity functions are defined as "1 − inhibition", i.e., high inhibition will result in normal activity ($H = 1$). The functions for residual activities are assumed to obey following equations.

$$H_1 = c_{11} H_1 - c_{12} H_1^2$$

$$H_2 = c_{21} H_2 - c_{22} H_2^2$$

The effects of 35 herbicides on the nitrification process were tested both by experiment, and by simulation of possible mechanisms of inhibition in a mathematical model by Domsch and Paul.[6] The model consists of nine equations with six coordinated constants and seven measurable parameters or initial values, depending on the specific soil. The first parameter studied with the help of the basic model was the effects of pesticides on the death rate of bacteria. Comparison of experimental to simulated experimental data did not fall closely to hypothetical curves. Similarly the effects of bacteria differed from hypothetical curves especially at the end of the nitrification process. The third hypothesis tested was based on the assumption that immediately after the application of pesticides a certain portion of the NH_4^+ oxidizers was killed and the surviving portion of bacteria proliferated normally. The results of this parameter come near to measured (Figure 5). A more satisfactory correspondence between measured and a simulated curves was obtained with a hypothesis of a completely reversible inhibition. In this case, the herbicide inhibited both enzyme synthesis and turnover rate of the NH_4^+ oxidation. In contrast to the previous hypothesis, the inhibition effects were completely reversible (changing over time). This means that in general, after a period of time, the oxidase system will return to a normal rate of substrate oxidation as the organisms regain the ability of normal proliferation (Figure 6). The simulated and measured dosage-response curve which corresponded to this hypothesis are shown in Figure 7. Another hypothesis which was found invalid in simulation runs was where soils with large initial population of NH_4^+ oxidizers were considered. In addition it was also considered that the surviving population should still be active enough to complete the process without appreciable delay.

Domsch and Paul[6] extended these studies to other than standard soils and it was observed that NH_4^+ oxidation was inhibited in fewer instances than expected. To explain these results, a hypothesis, was tested to determine if the population density had to reach a threshold value before significant inhibition of NH_4^+ oxidation occurred. According to this hypothesis soils with large initial populations of NH_4^+ oxidizers should be influenced, however, the surviving

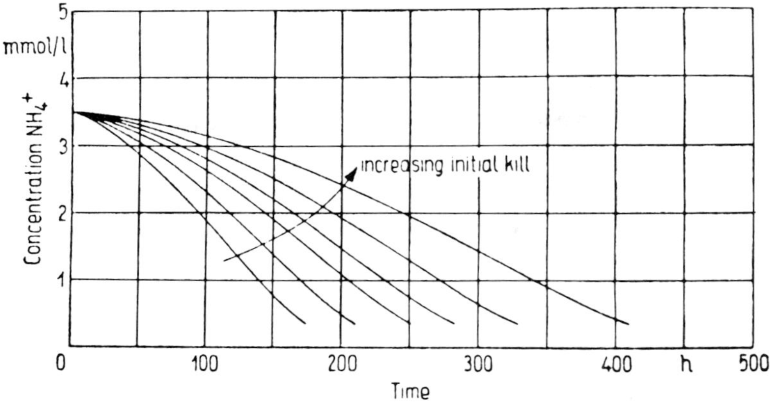

FIGURE 5.   Hypothetical curve for the NH$_4^+$ concentration assuming initial killing effects on the original populations E$_1$, (0) (= initial inhibition). The original population was varied between 100 and 20%. (From Domsch, K. H. and Paul, W., *Arch. Microbiol.*, 97, 283—301, 1974. With permission.)

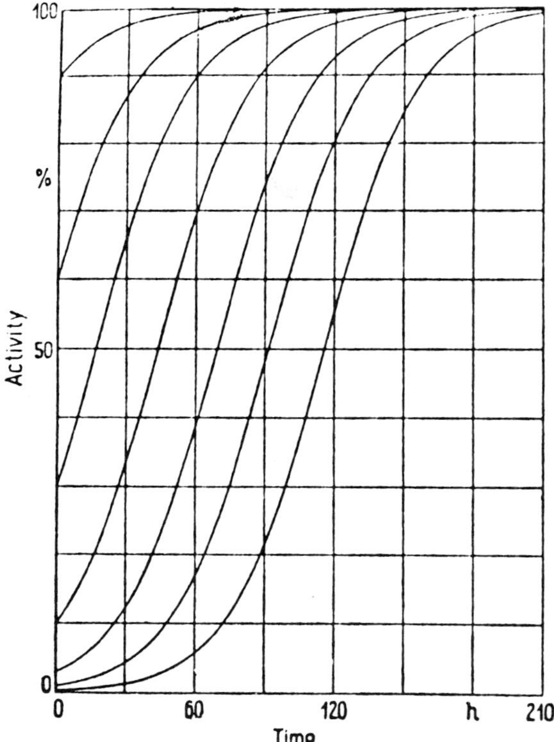

FIGURE 6.   Logistic activity time curve for the assumption of time dependent reactivations of the enzyme system (= completely reversible inhibition). (From Domsch, K. H. and Paul, W., *Arch. Microbiol.*, 97, 283—310, 1974. With permission.)

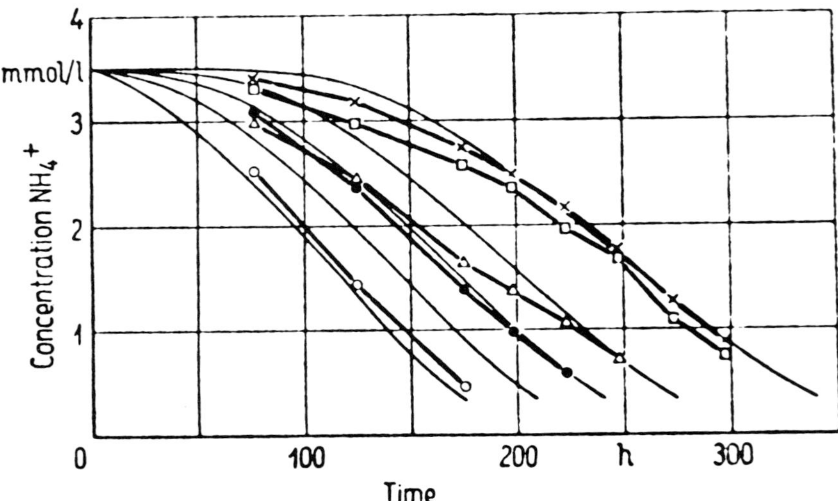

FIGURE 7.   Hypothetical curve (fine-lined) and measured curves (heavy lined) for the NH_4^+ concentration assuming a completely reversible inhibition in soil. I. Examples given: O——O control; ●——● barban (0.5 kg/ha); △——△ methabenzthiazuron (5 kg/ha); □——□ metobromuron (1 kg/ha); x——x diallate (5 kg/ha). (From Domsch, K. H. and Paul, W., *Arch. Microbiol.*, 97, 283—301, 1974. With permission.)

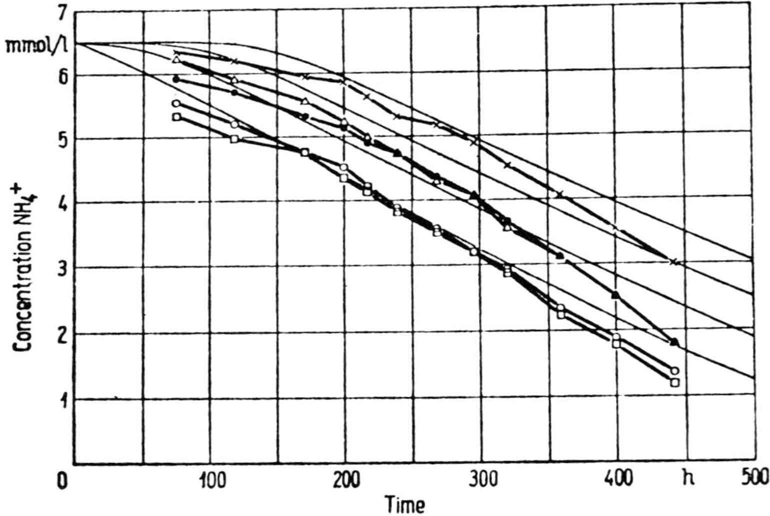

FIGURE 8.   Hypothetical curves (fine lined) and measured curves (heavy lined) for the NH_4^+ concentration assuming a completely reversible inhibition in soil P II (Pseudogley-parabraunerde). Example given; O——O control; ●——● barban (2 kg/ha); x——x metobromuron (2 kg/ha); □——□ diallate (5 kg/ha); △——△ methabenzthiazuron (5 kg/ha). (From Domsch, K. H. and Paul, W., *Arch. Microbiol.*, 97, 283—301, 1974. With permission.)

population should still be active enough to complete the process without delay. This hypothesis was also found invalid in simulation runs. Significant inhibitory effects became evident in soils only when pH in the soil fell below 7 (Figures 7 and 8). Analysis of experimental and model results gave strong indication that the conditions like pH of the soil for herbicide sensitivity of NH_4^+ oxidation were in complete agreement with experimental

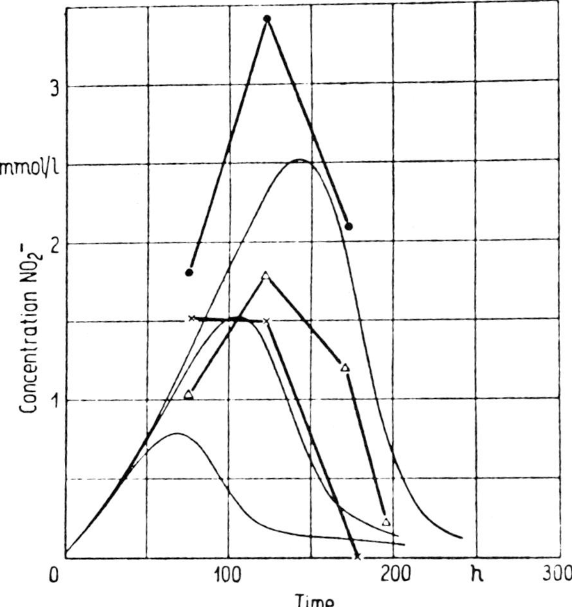

FIGURE 9.　Hypothetical curves (fine lined) and measured curves (heavy lined) of a short-termed NO_2^- accumulation assuming a completely reversible inhibition of E_2 in soil PPI (standard soil). Examples given; x———x triallate (2 kg/ha); △———△ diuron (2 kg/ha); ○———○ chlorpropham (5 kg/ha). (From Domsch, K. H. and Paul, W., *Arch. Microbiol.*, 97, 283—301, 1974. With permission.)

data on nitrifiers. It is thus evident that (using NH_4^+ oxidation as an example) the hypothesis of completely reversible or partially reversible inhibition comes nearest to measured values. In addition, Domsch and Paul[6] found that the second oxidation step of nitrification, i.e., the oxidation of NO_2^- ions, was more sensitive to pesticides (Figure 9); as a result, time limited accumulation of NO_2^- occurred. They also considered delay of NO_2^- flow by reversible inhibition of the first nitrification step. The second nitrification step was also found to be reversibly inhibited which was either complete or partial (Figure 10). The hypothetical and simulated values for inhibition of NO_2^- oxidation were nearer for the pesticides barban, methabenzthiazin, chloroxuron, linuron, and monolinuron (Figure 11).

This model includes H_1 and H_2 which take into account the inhibitory effects of the herbicides which were not considered in the previous models.[34-36] It is further distinct in the monod-type growth functions, the pH dependence of the optimal growth function, and distinction between free and adsorbed NH_4^+ ions are also considered. However, the validity of this model requires that no gradient down the soil column should occur. Thus, by numerous simulation runs with this model various hypotheses of the mechanisms and the degree of pesticide inhibition could be checked against the experimentally measured values. Readily demonstrated are the time dependence of inhibitory effects, the dependence on the pH-value, and different effects on *Nitrosomonas* and *Nitrobacter*. This model was also able to relate previously published work on many herbicides.

V. DENITRIFICATION

Denitrification in soil is usually closely coupled to other phases of the nitrogen cycle such as nitrogen fixation, ammonification, and nitrification and it is difficult to evaluate denitri-

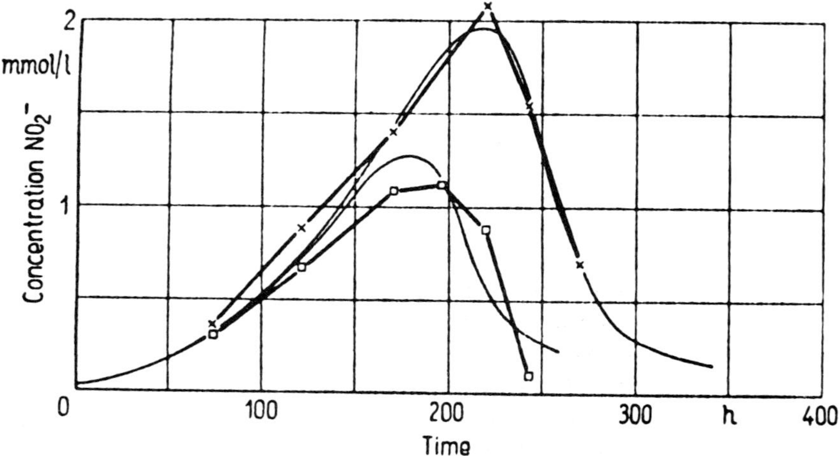

FIGURE 10.   Hypothetical curves (fine lined) and measured curves (heavy lined) of a short-termed delayed NO_2^- accumulation assuming a completely reversible inhibition of E_1 and E_2 in soil PPI. Example given; □———□ metobromuron (1 kg/ha); x———x (metobromuron (2 kg/ha). (From Domsch, K. H. and Paul, W., *Arch. Microbiol.*, 97, 283—301, 1974. With permission.)

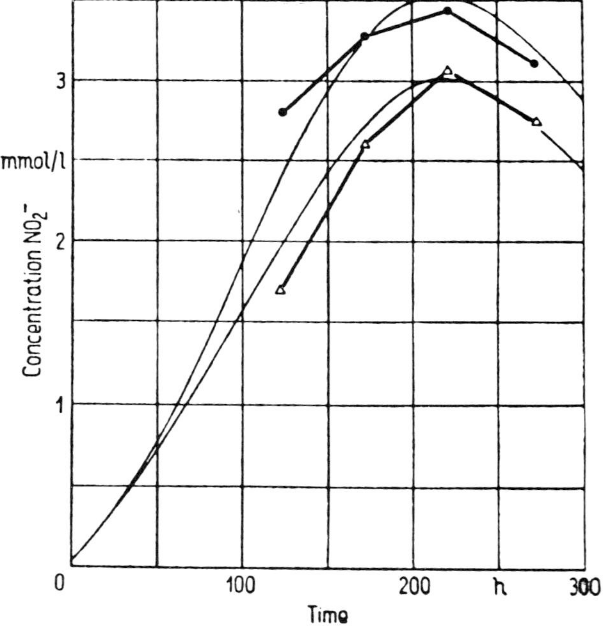

FIGURE 11.   Hypothetical curves (fine lined) and measured curves (heavy lined) of NO_2^- accumulation assuming a partially reversible inhibition of E_2 in the soil PPI. Examples given; △———△ meth-abenzithiazuron (5 kg/ha); ●———● barban (2 kg/ha). (From Domsch, K. H. and Paul, W., *Arch. Microbiol.*, 97, 283—301, 1974. With permission.)

fication in the presence of other gas evolving processes such as respiration. One method for evaluating denitrification is to inoculate two sets of tubes containing a liquid of nitrate and glucose which serve as the nitrogen and carbon sources respectively. The tubes of media are inoculated with aliquots of a serial dilution of soil and tests are made daily to check the disappearance of nitrates and the appearance/disappearance of nitrites. Using the most probable number method it is possible to calculate the number of responsible microorganisms and to construct an activity curve.[37] However, Todd and Nuner[38] found that the approach gave inconsistent results.

Analysis for the total nitrogen remaining at the end of the experiment can also be used to evaluate the losses in gaseous form. Wheeler[39] used the Lees-Quastel type of soil perfusion apparatus to study the relation between ammonification, nitrification, and denitrification, while McGarity[40] used a Warburg apparatus to trace the release of carbon dioxide, nitrogen, and nitrogen oxide in an argon atmosphere. A sophisticated apparatus was used by McGarity et al.[41] for continuous monitoring of the soil's humidity and oxygen content, the rate of oxygen consumed, and release of gases from the soil. Wijler and Delwiche[42] used a nitrate ^{15}N-enriched atmosphere to study denitrification. The isotopic distribution in the total nitrogen present as nitrate, nitrite, ammonia, atmospheric N_2, and amide-nitrogen was determined.

In the beginning for the bacterial taxonomist the ability of pure cultures to reduce nitrate and release gas into an inverted vial (Durham fermentation tube) or an unswept tube (Smith tube) served then the method to reveal the ability of microorganisms for denitrification.[43] In fact gas trapping is coupled with a MPN procedure for the quantitative estimations of the microbes involved in this process.[44-46] Alternatively, attempts have also been made to obtain the direct viable counts of denitrifiers.[47] To accomplish such enumeration, diluted samples are spread on the sterile surface of suitably nutritive agar in petri dishes. The agar cultures are then inverted and incubated under appropriate conditions until colonies are formed. The colonies are counted, picked to pure culture, and from there to nitrate broth to be tested for gas production. Choosing the appropriate components for the isolation medium, the best temperature for incubation, and the proper initial concentration of nitrate are critical to the success of such an approach.

Another approach to study the denitrification is to examine the components of the second half of denitrification, i.e., nitric oxide, nitrous oxide, and nitrogen. It was around World War II that accurately calibrated Bar-Croft-Warburg constant-volume respirometers became available to study denitrification.[48] This apparatus consists of trapping mechanisms for gases other than denitrogen, in liquids in the center well of the respirometer flasks and the output of denitrogen is measured by differential pressure.

The acceptibility of a variety of oxidizable substrate as electron donors, coenzymes as intermediate carriers, nitrate and nitrite as electron acceptors, and nitrogen oxides as probable intermediates made the analysis of denitrification more easy.[49-52] This aspect of trapping gases released by denitrifiers is known as manometry which is not always convincing as it is necessary to ensure that nitrogen gas as a result of dinitrogen is ultimately the gas that moved the level of the manometer fluid. This difficulty was solved with the development of gas chromatography techniques in 1966 when polyaromatic beads as (Porapak Q) packing material became available.[53] The methods allowed the separation of intermediate products of denitrification such as nitric oxide, nitrogen oxide, and nitrogen with great ease. Quantities of gases in micromole range are detectable. The occasionally encountered contaminants, oxygen and carbon dioxide are also separated from all the other gases.

Another packing material, Molecular Sieve 5A proved particularly useful for trapping nitrous oxide while permitting other gases of interest to pass through.[54] Large quantities of nitrous oxide are first allowed to accumulate on Molecular Sieve 5A and can then be flushed from the trap onto a Porapak Q column and assayed alone. This two-step procedure reduced the possibility of interference from other gases. Gas chromatography is now a widely used procedure for analysis of denitrification, not only of the activity in whole cells and extracts but also in field situation.

In addition to gas chromatography, compounds labeled with heavy, stable isotope ^{15}N, have also been employed to study the process of denitrification.[55-63] Most of the investigations, however, are restricted to ^{15}N-labeled nitrate only. Heavy nitrate is now being extensively used for analysis of denitrifying activity in agronomic and ecological experiments. Chromatographic separation of the component gases prior to the injection of test sample containing ^{15}N-labeled nitrate into the mass spectrometer permits qualitative estimation of ^{15}N-labeled material among other compounds released.

Many factors such as organic carbon, soil pH, temperature, nitrate concentration, water content, and oxygen concentrations have been elucidated as affecting denitrification.[64-66] However, the predictability and the determination of absolute amounts and rates of denitrification have met with only limited success. This has been due primarily to the lack of methods for quantifying gaseous flux occurring in soils and to the very complex interactions of all factors affecting biological populations and the denitrification process. A better understanding of the effects of pesticides on denitrification require not only a thorough understanding of the process and the application of methods for quantifying denitrification but also means of evaluating various environmental factors.

Very few experiments have evaluated the absolute amounts and rate of denitrification in the field. Rolston and co-workers[67] demonstrated that the volatile gases from denitrification could be measured in the field profile using ^{15}N. Total denitrification was determined by integrating with time the flux of the gaseous denitrification products, as determined from measured soil gaseous diffusion coefficient and concentration gradients. However, these studies only evaluated the amount of denitrification under one cropping system and one soil water content near saturation. Rolston et al.[68] directly measured the flux ^{15}N$_2$ and N$_2$O by placing covers over the soil surface and sampling the atmosphere beneath the covers after 1 or 2 hr. The method for correcting the gaseous flux was made from measurement of ^{15}N$_2$ and N$_2$O concentrations within the soil and the final ^{15}N$_2$ and N$_2$O concentration under the cover at the end of 1 or 2 hr. The N$_2$ and N$_2$O flux from field profiles was measured at high and low temperatures, at two water contents near saturation, and at three levels of carbon input (cropped, uncropped, and manure). The research demonstrates that denitrification occurs very quickly with extremely high rates under some situations in the field profile, that the addition of carbon through manure, and the presence of a crop growing on the soil greatly increases denitrification compared with that of a crop growing on the soil. The denitrifications from uncropped plots was very small regardless of how wet the profiles were kept. It was found that denitrification occurred over a very narrow range in soil water content or soil water potential and that denitrification occurred mostly in the upper 45 cm in alluvial well-drained loam soil.

The research by Rolston and co-workers[67] was conducted at constant soil water contents. However, the normal wetting and drying cycle which would take place under field situations through either rainfall or irrigation may drastically change the rates and total amounts of denitrification that may occur. The rate at which a microbial population can increase from a relatively small biomass of a dry soil to a population which effectively reduce nitrate is not known for a field soil. It is conceivable that a soil irrigated frequently would be relatively dry in the upper part of the profile with a low microbial population. During the irrigation cycle it may take several hours before a sufficient population could develop, yet soil water redistribution and decreasing water contents would begin shortly after irrigation. It might be expected that denitrification under such an infrequent microbial schedule would be small. However, it is known that a rapid increase in microbial activity occurs after dry soils have been wetted because of a readily available carbon source.[69] If the water is maintained on the soil for a long period an anoxic condition develops in microsites, and large denitrification

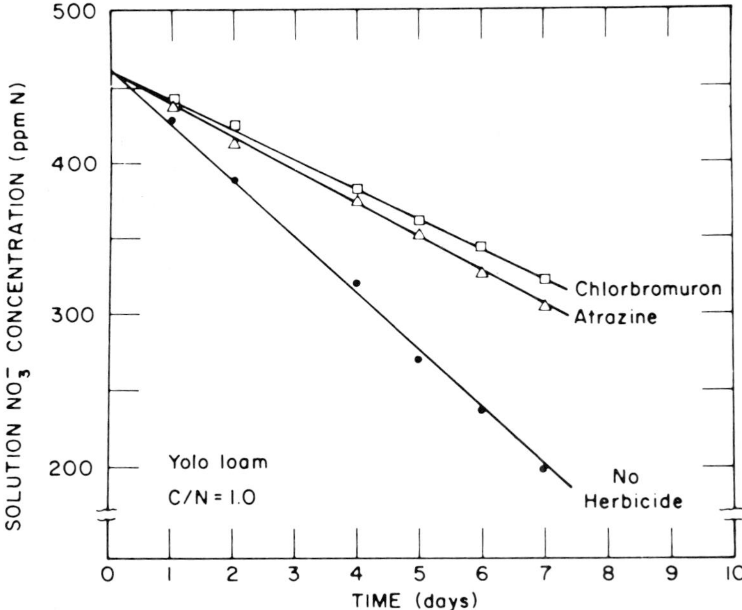

FIGURE 12.   Concentration of NO$_3^-$ solution as a function of time of incubation of yolo soil with no herbicide, chlorbromuron, or atrazine. (From Rolston, D. E. and Cervelli, S., in *Agrochemical Residue-Biota Interactions in Soil and Aquatic Ecosystems,* IAEA, Vienna, 1980, 189—199. With permission.)

rates may be possible with infrequent, large irrigation. For frequent small irrigations, a profile would remain wetter than that which would occur between infrequent irrigations, and this may result in a fairly constant microbial population in the upper portion of the soil profile. However, the amount of denitrification that would occur under this system might be either large or small dependent upon whether soil water contents became enough to result in the development of anoxic conditions in the soil. In such irrigated soils the disappearance of NO$_3^-$ is rapid with time (Figures 12, 13). Thus, while designing experiments on the effects of pesticides on denitrification all these factors must be given priority. Further in such soils the concentration of NO^2 at the time of pesticide treatment should be high enough.[70]

VI. ENUMERATION OF FREE-LIVING NITROGEN FIXERS

Despite its ambiguity, the plate technique for *Azotobacter,* devised by Winogradsky, remains a quick and reliable method for demonstrating the presence of aerobic, free living nitrogen-fixing bacteria. The method consists of mixing a carbon source (preferably mannitol) into soil moistened to a paste, which is then placed in petri dishes and the surface of the soil carefully smoothed over. After incubation, *Azotobacter* colonies can be seen on the surface of the soil. For a large number of soil samples the method can be used to obtain an approximate idea of the number of viable *Azotobacter* cells existing in the soil. It is preferable for examining the effects of pesticides on *Azotobacter,* as the test is conducted in soil. conducted in soil.

Dilution agar-plate counts using media to favor the growth of azotobacters is also subject to some difficulties and open to the same criticism as are total microbial counts. A less accurate but more practical method involves the use of the selective saline liquid media and dilution of soil samples using replicate tubes for each dilution.[71] The presence of *Azotobacter* in a tube after incubation is indicated by a characteristic film on the surface of the liquid

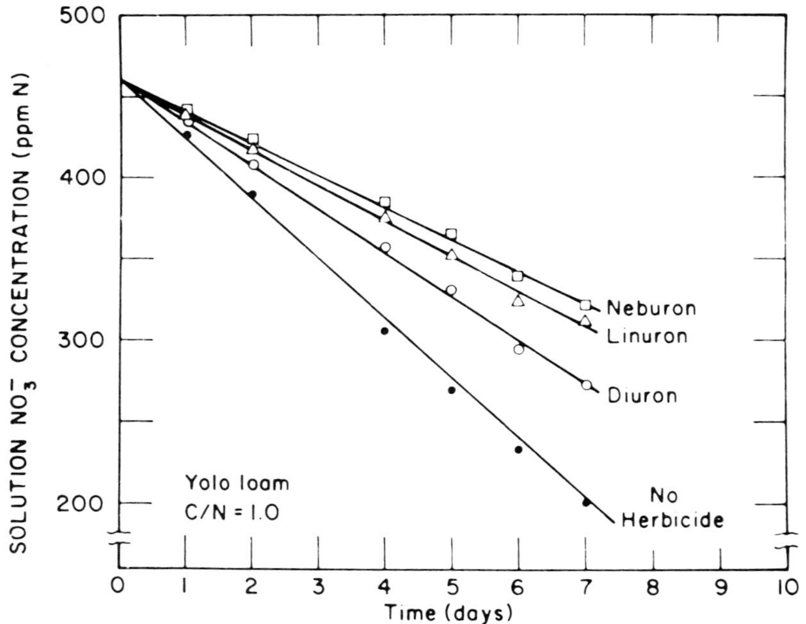

FIGURE 13.   Concentration of NO_3^- in solution as a function of time of incubation of yolo soil with no herbicide, neburon, linuron, or diuron. (From Rolston, D. E. and Cervelli, S., in *Agrochemical Residue Biota Interactions in Soil and Aquatic Ecosystems*, IAEA, Vienna, 1980, 189—199. With permission.)

medium. It is frequently necessary to confirm these observations by microscopy. Although more practical, this method is not ideal for studying the effects of pesticides since the process of dilution may remove nonlethal inhibitory effects of the pesticides. Incorporating pesticide into the medium so that all tubes contain the same amount of pesticide cannot be recommended since as dilution increases, the number of cells decreases, and effects of pesticides are exaggerated relative to undiluted soil.

A. Rhizosphere

The studies on the rhizosphere are more difficult than nonrhizosphere studies since the plant-root imposes additional parameters which influence the distribution, type, and response of the soil's microorganisms. Dilution agar plate counts become complicated because it is difficult to remove organisms from the root surface as well as the soil adhering to it. The nature and type of organisms involved in nitrogen transformations in the rhizosphere have already been described in detail in Chapter 1.

Estimates of the microbial number in rhizosphere soil can be made by dilution agar plate counts using a wide variety of media. Numbers of organisms developing on the plate can then be related to the weight of rhizosphere soil determined by the method of Timonin[72] or by the volume displacement method of Reyes and Mitchell.[73] A possible advantage of the method is that the amount of pesticide present in the rhizosphere soil can be determined by chemical analysis of the soil.

A root maceration technique in which rhizosphere soil is first removed by shaking roots in water after which the roots are removed and shaken in three changes of sterile water was used by Cook and Lochead.[74] The roots are then weighed and resuspended in sterile water in which they are macerated. Dilutions are prepared from the blended suspension and plated in agar media. Numbers of organisms are related to the weight of macerated root tissue.

Apart from the criticisms of dilution agar plate counts, a further disadvantage of this technique may be the inhibition of rhizoplane organisms by toxics such as phenols and alkaloids released from the plant. Variations of the root maceration technique are described by Stover and Waite[75] and also by Singh.[76]

Apart from dilution agar plate counts or agar plate isolation methods for studying rhizosphere organisms, several methods originally designed for nonrhizosphere soil studies have also been used to study the rhizosphere microorganisms. Thus, Starkey[77] adapted the Rossi and Cholodny buried slide method to study rhizosphere organisms. In this method microscope slides are buried vertically or horizontally at different depths in the soil. Seeds are planted above the slides and after root development has occurred, the slides are removed, dried, and stained in phenolic rose bengal and examined microscopically.

Parkinson's soil box is also another versatile device leading itself to isolation of organisms of the rhizosphere and their microscopy.[78] The box of the wood or perspex can be fitted on one side with removable perspex or glass microscope panels. The box is then filled with soil in which seeds are planted and is turned so that the plant roots grow down into the perspex or the glass panels. The box is then reoriented, so that the roots grow down the profile of the panel with perforations through which small tubes containing agar can be introduced such that the agar comes into contact with the root but not the soil. The tubes of agar can be removed at intervals, the agar core exuded, segmented, and the segments plated in a nutrient medium. If microscope slides are used in place of perforated perspex panels, these can be removed after marking the outline of the root on the back of the slide and stained, mounted, and examined microscopically for organisms adhering to the portion of the slide formerly in touch with the root. If the slides are not removed, fungal and bacterial cells attached to the roots can be observed through the glass slide by using a form of incident lighting.[79]

In general, all methods described above can provide estimation of the total biomass but cannot give an idea of the number of organisms of a particular group. In that case selective media can only impart information on specific activity of a particular group of organisms.

Microorganisms growing on or near plant roots can be counted by dilution plate or miniaturized MPN techniques as described previously. The methods differ only in the preparation of the initial suspensions from which the dilution series are prepared. With plants grown in the field the roots are lifted from the soil with a garden fork. When plants are grown in pots the entire contents of the pot are tipped out. In both cases the roots are teased from the soil mass and any large lumps of soil gently crushed with fingers and removed. At this stage any soil remaining in the root systems can be regarded as rhizosphere soil. This procedure works most satisfactorily in most soils at about 60% moisture holding capacity but may have to be varied slightly in very dry or especially dry/wet soils. In wet soils, simple shaking may not remove much soil, thus vigorous treatment may be required to remove nonrhizosphere soil. In general the greatest rhizosphere effect is considered to extend to only 1 or 2 mm from the root surface. In dry soils shaking may remove too much soil to allow a big enough sample of adhering soil to be obtained from the root.

Portions of root, normally primary, laterals, or entire seedling root systems, are placed in 100 mℓ of sterile water in a conical flask. The flask is closed with a bung and shaken vigorously. The resulting soil suspension is serially diluted and the microorganisms counted as described previously. The weight of the rhizosphere soil obtained is found after removing the root material from the suspension using a wire loop. The suspension is then dried in an evaporating basin on a boiling water bath. The dried soil is cooled in a descicator and weighed. If required the dilutions of this suspension can be combined and dried down with it, but in practice it is usually satisfactory to mathematically correct the dry weight of soil in the initial suspension for the amount removed in preparation of the dilution series.

B. Root Surface

Roots which have been washed free of rhizosphere soil are transferred from the flask to a sterile aluminum screwcap weighing tube lined with filter paper and weighed. Gentle shaking in this type removes much of the surface moisture held by the roots. The roots are then transferred to 250 mℓ flasks containing 100 mℓ of sterile water and 4 g of glass beads. The empty tube is weighed and the weight of the roots found by difference. The flask with the glass beads and roots is closed with a sterile rubber bung and shaken vigorously. The abrasive action of the beads serves to remove organisms adhering to the surface of the roots. The resulting suspension is diluted serially and the numbers of organisms present estimated.

C. Root Interior

Microorganisms colonizing the tissues of the root can be estimated in suspension prepared by macerating the root. Roots are removed from the flask containing glass beads with a wire loop and transferred to a sterile Cryobac bag after having weighed them in sterile filter paper-lined aluminum tubes. Sterile water (100 mℓ) is added to the bag and the roots are then macerated in the stomacher for 5 min. The resulting suspension is diluted serially and the numbers of organisms estimated.

VII. *RHIZOBIUM*-LEGUME ASSOCIATIONS

Rhizobium-legume association involves a complicated system of interactions. It is difficult to study the affect of pesticides on the entire process and for this reason in most of the studies either the *Rhizobium* alone or legume has been considered. A few studies have considered the effect on both the partners. Following are the methods which have been generally used to study the effects of pesticides on rhizobial growth, legume growth, yield and nodulation.

A. Enumeration and Determination of the Growth of Rhizobia

The colonies of rhizobia are generally raised in yeast extract mannitol medium. In this medium faster growing forms produce medium or large colonies after 3 to 5 days at 25 to 28°C. Colonies of slow growing rhizobia are barely detectable after 3 to 5 days and even after 10 days will commonly produce colonies not exceeding 1 mm. Ambient growth and marked pH change on peptone agar after 1 to 2 days at 30°C, rapid change of litmus milk at 26°C, abundant growth, and discoloration of potato slopes all argue against the culture being *Rhizobium*.

Complex media, commonly based on yeast extract, are also generally used for routine purposes. These and defined media for specific investigations, have been described in various places.[80] It is important with complex media to avoid the common practice of using too much yeast extract which results in inhibited growth or abnormal morphology.

Distinction between fast- and slow-growing rhizobia is often important both in the examination of an existing culture and in the process of isolation from a nodule. This requires continuous incubation and careful observations up to at least 10 days with inoculum so spread on the plate or on successive plates, as to secure well isolated colonies. Smaller colony forms of fast growers produce their colonies about as fast as the large, even though they fail to reach the larger diameter characteristics of the latter. In addition, ability to change the pH of yeast extract mannitol agar under well-aerated conditions, such as on the surface of a poured plate, is also a useful basis of distinction between the two forms. Under these conditions fast growers characteristically drop the pH sufficiently to change the indicator color from thymol blue to yellow, whereas slow-growers either fail to change the indicator color or move it in direction of alkalinity.

There is seldom any difficulty in obtaining abundant rhizobial growth from freshly collected nodules of healthy plants. Nodules from old plants and those taken from plants some time after sampling are kept in a moist atmosphere preferably under cool conditions. Deep freezing but not ordinary refrigeration permits nodulated roots to be stored for a long period. A choice between simple and a more elaborate procedure is determined by the nature of work being undertaken and facilities available. It should be noted that when several nodules are to be sampled from one plant, it is practical and much more convenient to handle an appropriate portion of the nodule-bearing root system in one operation.

Nodule contents can be conveniently extruded by squashing the nodule on the under surface of the lid of the petri dish containing the ready poured yeast extract mannitol agar. Then a small amount is streaked over the agar surface in a pattern likely to result in part of the spread plate having well-isolated colonies. In some cases the inoculum may be taken over and spread on a second prepared plate without recharging the inoculating loop. When it seems necessary, the inner part of larger nodules may be dissected out aseptically to lessen the likelihood of a mixture with persistent surface forms. The streaked plates are examined after incubation of 25 to 28°C for a shorter and longer period. Careful distinction should be made between colonies typical of rhizobia and other forms and between fast- and slow-growers. A colony, without contamination is used to establish a slope culture. Isolation from soil where other bacteria are likely to swamp out rhizobia on growth media, generally involve the use of legume as "trap" host.

The susceptibility of *Rhizobium* spp. to pesticides has been extensively investigated by a variety of methods ranging from colony counts on specific agar media to determination of leghemoglobin content and the ability of *Rhizobium* spp. to infect appropriate legumes. In parts of the world where leguminous crops are agronomic (especially in times of nitrogenous fertilizer shortages or intensification of agriculture), the effects of pesticides on the rhizobia, legume growth, yield, nodulation, total nitrogen content, and nitrogen fixation becomes a meaningful and practical area of study. However, there are practical difficulties which one encounters while studying the effects on rhizobia-legume associations. Though the need for the development of a media that can facilitate the isolation and enumeration of soil rhizobia has long been recognized, the prospects for the development of such media are very poor. Physiological diversities of both rhizobia and soil microflora pose certain serious problems so that a medium to inhibit all other bacteria (yet be completely selective for a wide range of rhizobia), is most unlikely. Some media are sufficiently selective for limited application as in the case of enumerating rhizobia present in abundance in previously sterilized peat[81] and for differentiating strains isolated from nodules.[82]

A medium containing antibiotics, pentachloronitrobenzene, and sulfafurazole was considered as a selective medium for soil isolation but careful and extensive work by Pattison and Skinner[83] demonstrated very clearly the difficulties in formulating a selective medium for the rhizobia. They tested 47 strains representing six species of *Rhizobium* and the cowpea for the sensitivity to six antibiotics. Each strain was inhibited by at least one antibiotic and no consistent patterns of inhibition were found within a species. The only antibiotic that had little effect on rhizobia was penicillin at a low concentration. When high concentration of penicillin was incorporated into a selective medium a strong inhibition of growth in most of the strains of rhizobia was noticed. Pattison and Skinner[83] tried 19 of their strains on the medium of Graham[84] and none grew. Thus, selective media adequate for the isolation and enumeration of rhizobia in soil or rhizosphere apparently are out of question except as a limited possibility for specific strains.

Not much attention has been given to development of techniques to study the rhizobia in soil. This is one reason why the ecology and subsequent effects of pesticides on rhizobia have not been studied extensively. Because of the lack of fundamental methods it has become further difficult to study the effects of pesticides on rhizobia-legume associations. In addition to the normal techniques which are commonly used to study the effects of pesticides on rhizobia, fluorescent antibody technique which is more recent has also been described. This technique has not been used to study the effect of pesticides on rhizobia but the possibilities of this technique becoming more useful for this type of work cannot be ruled out.

1. Total Growth

While studying the effects of pesticides on the growth of rhizobia, normal growth enumeration methods are employed. To the test material appropriate concentrations of pesticides are added and untreated control is kept simultaneously. At regular intervals of time the growth of the organisms is monitored. The choice of methods for the enumeration of the effects on rhizobia depends on the nature of the material and the objective of such measurements.

If we are interested in studying the effects of pesticides on the total number of rhizobia (both viable and nonviable) then we do a total count: either directly under the microscope or indirectly by determining turbidity and applying the appropriate conversion factor. The latter method can also be related to total cell substance. However, the turbidity method is not applicable to pesticides which are insoluble in water as a pesticide beyond solubility will precipitate and interfere with the absorbance measurements.

When our interest is in viable count we have to use a counting method which tests ability of the cell to grow on a suitable medium, as in the plate count. To this we may add the requirement of being able to nodulate on appropriate legume as in plant infection count. The latter method is particularly useful when the rhizobia are in minority amongst other microorganisms in soil or inoculants made up of a nonsterile media. Details of these procedures are given in practical manual of Vincent.[80]

a. Direct Microscopic Method

Direct microscopic method requires the use of clean slides, cover slips, and a deeper well (0.01 cm) of hemocytometer if the special bacteria counter (0.001 cm) is not procurable. Phase microscopy, with careful adjustment of light source, condenser, and phase rings is better than staining procedure. Movement of suspended bacteria during counting can be a problem. Inherent mobility can be inactivated by mild heat, counting needs to be done systematically and with critical assessment of what constitutes a cell. This is very important because treatment of pesticides with low solubility may form precipitates which can easily be mistaken for bacteria. In addition the standard error becomes proportionately less as the mean number increases, it is desirable to count as many cells as possible.

b. Turbidity

This method is convenient for progressive measurement of growth, if cultivation is in glassware having matched optical paths so that reading can be taken without actual sampling. This method is, however, relatively insensitive, so that according to the nature of the instrument and the method used, cultures are likely to require 10^8 and 10^7 cells per milliliter. In addition this method is also a poor predictor of viable numbers.

2. Viable Counts

Though this method is superior to microscopy and turbidity methods, its accuracy is also dependent upon many factors. These factors are full dispersion of the rhizobial cells, maintenance of viability, avoidance of growth during the total operation, sufficiently uniform distribution through the plating medium, and ability of the medium and growth conditions to permit colony development from each viable cell. The procedural details have been described by Vincent.[85]

3. Plant Dilution or Plant Infection Techniques

The selective medium most widely used to detect the presence of rhizobia in soil has been the legume host itself. The method is indirect as it depends upon the eventual appearance of a nodule on the roots of a legume test plant as the indicator of an appropriate *Rhizobium* in soil. Essentially, the method consists of inoculating test plants and growing them aseptically with aliquots from a dilution series of the sample under examination. The number of rhizobia in the sample can be calculated from the proportion of test plants forming nodules at each dilution. The quantitative accuracy of plant infection test depends on the ability of a single rhizobial cell to initiate nodulation of the test plant. This assumption appears valid provided the test plant is grown on agar in cotton wool, plugged test tubes. However, data from experiments in which test plants are grown on vermiculite have shown that the number of nodulated test plants are always less than expected indicating that a single rhizobial cell is not always capable of producing nodules on a test plant growing in a tube of vermiculite.

a. Soil Sampling

As the size of field populations of rhizobia may vary from point to point even over a short distance, considerable care should be taken to obtain a uniform sample. The area to be sampled is divided by eye into a grid of 25 sections.[85] A subsample of soil, roughly cylindrical in shape, 9 to 10 cm deep and 3.5 cm in diameter, is taken from each section, using a presterilized coring implement or a long-nosed trowel, and sealed in a polyethylene bag. Because of the dynamic nature of populations of rhizobia in soils, soil samples should be examined immediately after collection. In the laboratory the subsamples are thoroughly mixed on a clean, sterilized, solid surface, quatered and mixed again until a small composite sample as homogeneous as feasible is obtained.

b. Selection of Test Plants

It is preferable to grow the test plants within test tubes plugged with cotton wool. When counting rhizobia in soil using test plants cultured in agar, prolific growth of algal and saprophytic fungi sometimes interferes with nodulation. Soil-borne fungi and algae can also be controlled by growing the test plant on a vermiculite substrate. Furthermore, certain species suitable for use as test plants grow better in vermiculite culture than in an agar medium.

c. Seed Preparation

Clean, undamaged seeds, selected for uniformity by weighing, hand sorting, or sieving are dipped momentarily in 95% ethanol and then immersed for 5 min in 0.1% $HgCl_2$ solution. They are washed thoroughly in not less than ten changes in sterile distilled water to remove all traces of $HgCl_2$. These seeds are allowed to stand in the final change of water for several hours until they are fully imbibed. They are then spread on to 2.0% agar in petri dishes and incubated at an appropriate temperature in an inverted position to provide uniform seedlings with straight radicles. When the radicles are 1 to 2 cm long, the seedlings are transferred aseptically into the test tubes (one per tube). With agar culture the seedlings are placed pointing down with the root (in particular the growing tip), in contact with the moist agar surface. Care is needed as the root is fragile and easily damaged. In vermiculite culture, the seedling is buried vertically most conveniently against the wall of the tube, with the seed coat approximately 2 mm beneath the surface of the vermiculite. After planting, it is necessary to lightly press the surface of the vermiculite with a sterile, flat ended tamp in order to provide a firm seed bed and uniform emergence of the seedlings. The tubes are then irrigated with sterile distilled water or dilute seedling solution and placed in a shaded glasshouse or a lighted growth room.

d. Plant Infection Test in Tubes

There are many versions of plant infection technique but the method for counting rhizobia described here is a modification[85] of the one used by Brockwell.[86] A 10 g soil sample is suspended and shaken at 200 to 300 c/min on a wrist-action shaker. Then 1 mℓ of this suspension is pipetted into a 4 mℓ amount of diluting solution and shaken for 5 min further. From this, four 1 mℓ aliquots are used to inoculate four test plants. The remaining 1 mℓ of solution is used to make the next fivefold dilution in the same way, but without shaking. This is repeated to provide six successive dilution and thus uses 25 test plants (5 plants are inoculated with the last dilution). The test plants are grown for 4 weeks after inoculation, watered when necessary, and are then examined for the presence of nodules. From the proportion of plants forming nodules at each dilution level, the most probable number is calculated using a modified version of the original sample of Fisher and Yates.[87]

The above given procedure is applicable for a fivefold dilution series and can be used if the number of rhizobia is slightly higher. However, a tenfold serial dilution plant infection test can be used when a lower level of accuracy is adequate. The method for preparation of the material under examination is similar to that used for fivefold dilution. From each of the six tenfold (1 mℓ into 9 mℓ) serial dilution, three 1 mℓ aliquots are used to inoculate three test plants: thus 18 plants are required for each series. As before, the most probable number of nodule bacteria in the original sample is calculated from the proportion of test plants forming nodules at each dilution level.[85]

Since the method is a large expensive of time, materials, and space, the number of replicate tubes (plants) per dilution is commonly reduced from the usual MPN protocol of 5 or 10 to 2 or 4, with corresponding decrease in accuracy. Calculations based on statistical tables are made with the assumption that a single *Rhizobium* can initiate nodulation. Most workers seem to have been guided by the plant-dilution protocols of Date and Vincent.[81]

Enumeration by the plant-dilution technique has been the basis of virtually all ecological studies dealing with the persistence of the legume nodule bacteria in soil and their response to rhizosphere conditions. Since other methods were not available it is unfortunate to find, as Date and Vincent[81] have observed, that there is little published evidence to assess the validity of the method. These authors are among the few to report the validity of MPN determination for the pure culture system. Vincent[88] and Brockwell[86] found that a number of kinds of rhizobia gave plant dilution counts in pure trials that were in good agreement with plate counts. Similar controls in a study by Tuzimura and Watanabe[89] were much more variable. Thompson and Vincent[90] made further evaluations of plant dilution estimations in the light of unexpected skip encountered in the dilution series. They suggested that plant infection method was applicable to soils with rhizobia at population densities greater than 100/g of soil. However, it is surprising that this method has not been widely used even in cases where soil contains rhizobia more than 100/g of soil. In this case the impact of the plant in the nodulation process is an important factor. Within the same species of rhiozobia the host plant may select a particular type of strains among the total strains of rhizobia present in a given soil.[91] For example, certain soybean genotypes are selective for a particular strain in *Rhizobium japonicum*.[92] Legumes other than soybeans are also highly selective of their rhizobia.[93,94]

4. Immunological Approach

Our understanding of the ecology of *Rhizobium* in soil has been based mainly on percentage nodulation data. Although this information is of practical importance it does not directly address the question of rhizobial ecology because nodule formation is the result of a number of interacting processes[95] that need not reflect changes in soil and rhizosphere populations of rhizobia. Until recently numerical estimation of soil population levels of specified bacteria has been difficult. The advent of immunofluorescence techniques[96-101] and antibiotic marker techniques[102-105] have made it possible to study the autecology of individual serogroups of *Rhizobium japonicum* directly in soil and in natural rhizosphere.

Both the fluorescent antibody and antibiotic techniques have been used extensively for nodule identification and recently these techniques have been also used to quantify rhizobial populations of both soil and rhizosphere.[98] These techniques have not been used for studying the effects of pesticides on rhizobia-legume associations but there is every possibility that they may become popular to study the effects of pesticides.[106-108]

Serological reactions of rhizobia generally involve two major reactions: antigen and antibody (or complex polysaccharide). The latter is produced by an animal in response to a foreign antigen. Specific conformity between antibody and antigen determinant permits recognition of identity or relatedness. Antibody occurs in several molecular forms which share the same specificity for antigen but differs in the case of demonstration. The forms of antibody most utilized in vitro are the large and small immunoglobulins. Balance between the two forms varies according to such factor as stage and nature of the immunization procedure; the nature and location of the antigen. Absorption of antibody from antiserum can be used to demonstrate specificity and antigenic identity and to provide monospecific antisera. Some reactions can occur without prior exposure to antigen or test substance. Antisera to some strains of pneumococcus can precipitate polysaccharide of diverse origin, including exopolysaccharides of some rhizobia. Some plant proteins (lectins) act as pseudoantibody and combine with cell surface components of some rhizobia.

Three forms of serological reactions have been widely used in the study of rhizobia. Two of them (agglutination and combination with fluorescein conjugated antibody) involve antigen available for a reaction at the surface. The third commonly demonstrated as precipitation of antigen-antibody complex in a gel diffusion system, detects a wide range of soluble or solubilized antigens — internal as well as on the surface.

In this technique antibodies are prepared against the microorganisms to be studied and after the addition of a fluorochrome, the antigen-antibody reaction is visualized by fluorescent microscopy to allow for the simultaneous detection and recognition of the microorganisms of interest. However, this technique also has certain limitations. The detection of rhizobia is specific at the strain level and only *Rhizobium japonicum* has been studied in detail.[109] However, this technique can also be extended to normal nonsterile soil environment.[109,110] A major problem in the application of immunofluorescence techniques to soil environment is the specific staining reaction. Problems due to nonspecific staining are solved to some extent with the development of gelatin preparation used to treat specimens prior to the addition of the fluorescent antibody (FA).[110] This technique further has only quantitative application and is not adequate to probe any autecological questions of a host that involve population changes. The main difficulty with quantification by direct microscopy stems from the limited area of the microscope field at the magnification needed to see the bacteria. Therefore, in order to work with natural populations and to observe population changes it is necessary to remove bacteria from interfering soil and concentrate them for enumeration. Suitable techniques for the release of rhizobia from soil and their concentration on black membrane filters for enumeration of FA stained cells have been outlined.[96,111] The lower limits for enumeration by quantitative FA are 10^3 to 10^4 which depend on the dispersion and flocculation properties of the soil.

Information on the ecology of *Rhizobium meliloti* is limited and is in contrast with the extensive literature on *Rhizobium trifolii*[112,113] and *Rhizobium japonicum*.[98,99] This can be attributed in part to the success achieved in developing serological methods of identification for these latter two species. There are few reports on the successful use of serological methods for identifying and differentiating *Rhizobium meliloti*.[114-117] The method was refined to further characterize *Rhizobium meliloti* isolates[115,118] and the stability of both the antigenic properties of the isolates. The antisera were observed over a 10 yr period.[117] Dudman[114] introduced the technique of immunodiffusion for differentiating two *Rhizobium meliloti* strains which possessed common antigenic determinants.

In recent years, several reports have been published which showed various problems with serological methods for the identification of *Rhizobium meliloti*. These problems have been attributed to the preparation of antigens for serological assays,[119,120] the loss of agglutinating specificity by stock cultures[121] that was attributed to the loss of strain specific antigens, and the variability in the antisera obtained from different rabbits in response to the same antigen.[122] The contrast between success and failure of serological methods for *Rhizobium meliloti* identification may be due in part to the seemingly random choice of isolates made by many workers.

Observations documented in literature emphasize that serological characteristics of *Rhizobium* strains are stable, in contrast with other phenotypic characteristics, such as symbiotic effectiveness.[116,117,123] However, in recent years workers have attempted to develop alternative methods of strain identification which are inexpensive and rapid, and with which a large number of isolates can be handled. The methods of intrinsic levels of antibiotic resistance,[124,125] antibiotic resistance markers,[126,127] and phage typing[128,129] have been used in an attempt to circumvent time consuming and expensive serological methods. Although justifiable criticism of each of these methods has been raised, serological methods remain the most useful for direct observation of rhizobia within soil and in nodules, particularly in situations where the indigenous *Rhizobium* populations are heterogenous.[130]

Fuquay et al.[131] used microslab modification[132] of one-dimensional sodium dodecyl sulfate-poly acrylamide gel electrophoresis for separating cellular proteins profile patterns of eight field and one commercial *Rhizobium meliloti* isolates in order to choose candidates that were either identical or distinctly different from each other for the production of antisera. Their studies provided a rationale for selecting isolates to which antisera could be raised and to appraise the suitability of published methods of preparing *Rhizobium* antigens for the serological identification of field isolates. Figure 14 shows the one-dimensional protein profile patterns of eight field isolates and one commercial inoculant strain of *Rhizobium meliloti*. On the basis of their agglutination liters and gel immunodiffusion analysis, the isolates were placed in five serogroups which were identical to the groupings based on protein profiles. Antigenic characteristics of gel immunodiffusion antigens were influenced by the composition of the growth medium, sonication of the whole cells (Figure 15), and the addition of formalin. Isolates 27 and 31 reacted identically to antiserum 27. The reciprocal reactions of the same antigens with antiserum 31 were also identical, as were reactions involving isolate 21. Isolate 8 and 17 reacted identically against antiserum. The reciprocal reaction between either antiserum 8 or 18 and the two isolates (8 and 18) were also identical. In contrast gel immunodiffusion reactions between two isolates representing each of the groups showed further evidence of their nonidentity. Isolate 17 and 27 reacted to antiserum 17 in different way. The data reported herein support the need for caution in the use of serological methods as previously documented.[120-122] Thus, there is a need to scrutinize the general serological performance of any new isolate used in the future to ensure that methods of antigen preparation do not complicate the results. Further SDA-PAGE should be considered a potential tool with which to screen a collection of *Rhizobium meliloti* isolates to provide a better guarantee of obtaining antisera representative of different types of organisms from the collection and with the possibility that some of these sera may possess specific agglutination and immunodiffusion characteristics.

B. Legume Growth, Yield, and Nodulation

These tests should be restricted to situations where symbiotic nitrogen fixation may be affected by the pesticides, i.e., where pesticides are used directly on legumes, where legumes are a normal constituent of the crop rotation system, or where mixed leguminous and nonleguminous cropping is practiced. The choice of the test legume should depend on the "use pattern" of that particular pesticide.

The assessment method should recognize the unique symbiotic relationship between host

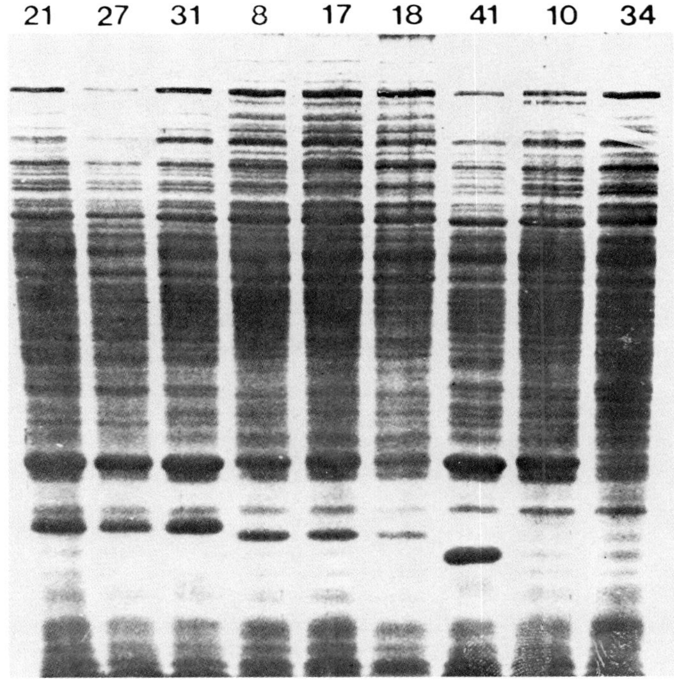

FIGURE 14.   Protein profile patterns of eight field isolates (21, 27, 31, 8, 17, 18, 41, and 10) and one commercial inoculant strain (34) of *Rhizobium meliloti* with SDS-PAGE. The gel is a single concentration of 11% (wt/vol) acrylamide. (From Fuquay, J. I., Bottomley, P. J., and Jenkins, M. B., *Appl. Environ. Microbiol.*, 47, 663—669, 1984. With permission.)

plant and rhizobia, and hence should determine in one test both plant and bacterial response to the pesticide. The following information will therefore be required: measurement of plant growth overtime, plant yield, and an estimate of healthy nodulation. The experiments should be conducted in the fields or in pots containing a soil type suitable for growth of the plant. The plant seeds would be inoculated with rhizobia only if no suitable bacteria are present naturally in soil. If the pesticide causes a significant effect on plant growth, further data should then be provided on the direct effect of the pesticides on *Rhizobium* organisms, the nodulation process, plant growth, yield, and the nitrogen fixation. Greaves et al.[130] described a method to test the effect on these parameters. Accordingly legume seeds are sown in 12.5 cm plant pots filled with a soil which is known to give full nodulation of the legume. The number of seeds planted vary according to the nature and size of the seeds (5 seeds of clover and 2 seeds of peas and beans). The pots are kept in a glass house with a day length of 12 hr (obtained using suitable fluorescent, mercury vapor, or sodium lamps in winter months). After germination the seedlings are thinned to 1 per pot leaving a stand of uniformly sized seedlings. Sufficient pots are prepared to allow 10 replicate plants to be taken at each sampling date. The pots to be treated with pesticides are sprayed when the legume has reached the stage of growth at which the manufacturer recommends treatment. Alternatively the pesticide can be uniformely incorporated into the soil by spraying with a laboratory pot sprayer. After spraying, the soil is mixed thoroughly. This soil is used to make upper 1.5 cm layer of the pot. The rates used are $1 \times$, $4 \times$, and $10 \times$ the manufacturer's recommended rate. Whenever experiments are to be performed in the field the legume seeds are sown

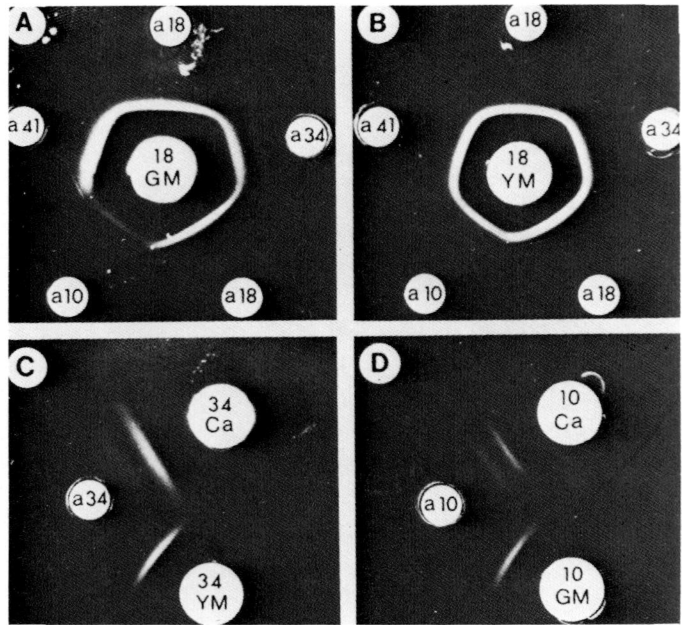

FIGURE 15. Immunodiffusion patterns of selected *Rhizobium meliloti* isolates illustrating the effect of medium for growth of cells on precipitin line formation. (A) isolate 18 grown in GM defined medium, sonicated; (B) isolate 18 grown in YM complex medium, sonicated; (C) isolate 34 grown either in GM medium supplemented with 0.5 mM $CaCl_2$(Ca) or YM medium, sonicated; (D) isolate 10 grown either in GM medium or GM medium supplemented with 0.5 mM $CaCl_2$, whole cells (A) Antiserum. (From Fquay, J. L., Bottomley, P. J., and Jenkins, M. B., *Appl. Environ. Microbiol.*, 47, 663—669, 1984. With permission.)

thinly in rows 60 cm apart. After germination seedlings are thinned out to 60 cm spacing within rows. This enables a single plant to be sampled whenever required.

In both pots and field, experiments plants are sampled at regular intervals. After spraying, 10 plants are taken from each treatment and from control. Plants grown in the field should be lifted from the ground using a spade, taking care to avoid damaging the roots. Plants are generally grown on the fruiting stage and sampled at 2 to 4 week intervals.

Numbers of nodules per root system are counted and the nodules are then removed from the roots and weighed. These are then dried at 80°C for 16 to 18 hr and reweighed. Roots and the aerial parts of the plants are weighed similarly. When fruits are formed they are weighed separately from the aerial parts to give a figure for final yield. All parts of the plants are analyzed separately for the total nitrogen (method described later).

Greaves et al.[130] highlighted the difficulties of designing experiments to assess the side effects of pesticides on legume and root growth, nodule and nitrogenase activity, and the problems of assessing the significance of observed effects. The difficulties were also highlighted by Johnen and Drew.[133] Greaves et al.[130] showed that large standard errors were associated with the results of plant growth, nodulation, acetylene reduction activity, and that relatively large coefficients of variation were obtained with measurement of nodule numbers, plant growth, and nitrogenase activity (Table 2). Thus, these methods would only detect major differences between treatments. Shoot, root, and total weights of the legume plants were less variable and could therefore, detect smaller differences. An additional problem of the lack of reproducibility of repeat experiments was encountered by Greaves

Table 2
EFFECTS OF ALLOXYDIM-SODIUM ON GROWTH, NODULATION, AND NITROGENASE ACTIVITY IN PEAS

	Application rate kg/ha	Experiments			
		1	**2**	**3**	**4**
Shoot dry wt g/plant	0	1.57	1.82	2.7	5.9
	2	1.34	1.12	2.8	5.8
	4	1.37	1.42	2.9	5.2
	SE±	0.08	0.13	0.2	0.3
Root dry wt g/plant	0	0.26	0.23	0.55	1.01
	2	0.20	0.16	0.67	1.27
	4	0.17	0.18	0.60	0.93
	SE±	0.02	0.01	0.04	0.13
Nodule dry wt mg/plant	0	6.6	7.1	83.5	217.1
	2	5.4	4.9	74.8	182.0
	4	2.9	4.3	50.9	259.4
	SE±	0.7	0.9	7.0	27.0
Nodule number/plant	0	41	38	301	317
	2	33	24	282	443
	4	24	36	233	429
	SE±	4	7	45	42
Nitrogenase activity nM N_2/hr/g root	0	185	6607	4053	2256
	2	143	5482	1688	4484
	4	116	6950	2576	2502
	SE±	18	557	233	504

Note: All the figures are the means of three replicate pots, six replicate plants.

From Greaves, M. P., Lockhart, L. A., and Rechardson, W. G., Proc. 1978 Br. Crop Prot. Conf. Weeds, 1978, 581—585. With permission.

et al.[130] They suggested that these problems were basically due to different means of measuring the side effects of pesticides on legumes. As the amount of labor on such experiments is high, this limits the number of replicates that can be handled at one time which in turn effects the precision of the results obtained. Further the analysis requires the services of 4 to 6 members and the nodule counting takes quite a long time. The measurement of nitrogenase activity also presents several problems.[130] As activity increases with increasing light intensity, the time of the year at which the plants are grown and the time of the day at which they are harvested, will effect the results. The use of growth chambers was suggested to maintain such controlled conditions in order to avoid these variations. In addition, soil conditions may also affect the design and interpretation of experiments. It is known that inorganic nitrogen inhibits nodulation and nitrogen fixation[134] and it was noticed by Greaves et al.[130] that peas receiving no fertilizer had about 100 nodules per plant at the time of spraying whereas those which were fertilized had none. Thus, the effects of pesticides may be of little importance where the crop is grown with fertilizer. Further, it is generally difficult to obtain uniform plants even in highly controlled experiments. The only way left under such a situation is to keep a larger number of replicates and to select only those plants (which appear almost equal) for treatment.

C. Nitrogen Fixation in Legumes

The three approaches used to measure N_2-fixation in legumes are based upon

1. Nodule number, mass, and leghemoglobin concentrations
2. N analysis
3. Acetylene reduction technique

The last mentioned method is described in detail in Section IX of this chapter.

1. Nodule Number, Mass, and Leghemoglobin Content

This class of measurements is the simplest but the least quantitative and with the possible exception of leghemoglobin concentration, none is recommended. The hematin content, determined as pyridine hemochromogen which consists largely of leghemoglobin parallels the amounts of N_2 fixed in peas, horse beans, and soybeans inoculated with different bacterial strains. Concentration, as opposed to total quantity of hematin in the nodules, provides the best estimates of N_2 fixation, but in soybeans this correlation decreases as vegetative growth ceases and the nodules turn green. The correlation between the hematin content and effective central-tissue volume of various host-plant nodules prompted the suggestion that the relation between N_2 fixation and hematin concentration was not of a cause-effect nature, but rather that the hematin concentration was an index of the volume of active tissue.[135]

When evaluated against alternative techniques now available for measuring N_2 fixation, leghemoglobin content only approximates quantitative measurements of N_2 fixation. Nodules formed by ineffective bacterial strains do not contain leghemoglobin and effective nodules lose their N_2-fixing capacity when the porphyrin ring of leghemoglobin opens. Methods and modifications of extracting hematin from legume nodules have been described and used.[135,136] The hematin generally is extracted in ammonium sulfate to prevent interference and denaturation of the pigment, and oxidation of iron by quinones formed by oxidation of phenolic substances during extraction. The hematin resulting from these extractions has been estimated spectroscopically as pyridine hemochromogen and upon treatment with an excess of sodium hyposulfite produces the deoxygenated form of hematin with an absorption maximum at 557 nm. Because the pyridine hemochromogen assay has disadvantages based on a sharp absorption peak, an unstable colored complex, and the irritating pyridine odor, an improved extraction procedure with Drabkin's solution and colorimetric measurement as cyanmetleghemoglobin was developed.[137] A new assay for hematin derivatives is 100 times more sensitive than the Drabkin's assay and permits replicate determinations on less than 1 mg of nodule tissue. In this assay iron is removed from hemin derivatives, converting them to highly fluorescent protoporphyrins that can be measured.

2. N Analysis Methods

The amount of N_2 symbiotically fixed can also be measured by Kjeldahl or 15_{N_2-} enrichment methods (described in detail later). However, these methods do not distinguish between N fixed in the plant from N_2 and N taken up from combined soil N. Therefore, in soybeans or other legumes suitable controls must be provided to correct for combined N uptake.

A correction method used in short-term greenhouse studies compared the total amount of N from combined sources and from N_2 in effectively nodulated plants with that in either inoculated with an ineffective bacterial strain or left uninoculated. In field experiments with peas, alfalfa, and soybeans, the controls employ low N fertility, low indigenous effective bacterial strain populations, uninoculated seed, seed inoculated with an ineffective bacterial strain, or a massive soil application of an ineffective bacterial strain to suppress the naturally occurring effective bacterial strains.[138] In such tests care must be taken to prevent contamination, and it is assumed that control and effectively nodulated plants take up equal amounts of combined N from the soil, an assumption likely to underestimate N_2 fixation.

Nitrogen fixation also has been estimated by the difference between total N of nodulating and non-nodulating isolines of soybeans.[139] However, the amount of soil N taken up by non-nodulating soybean isolines with their greater root proliferation[139] may not be comparable to that taken up by nodulating plants that are presumably less dependent upon soil N.

Another approach uses the difference in total N between a nodulated legume and a nonlegume crop, generally a cereal crop with a similar soil N requirement as uninoculated legume.[95] This method is simple but is not absolutely quantitative since under some conditions the nonlegume does not remove equivalent amounts of soil N as the uninoculated legume.

Several possible nonfixing plants could be used for soybeans.[138]

1. Nonnodulating isoline
2. Ineffectively inoculated nodulating line
3. Uninoculated nodulating line
4. Non-legume

The ideal prerequisites of the nonfixing control plant are that it should (first) assimilate its soil and fertilizer in identical proportion (but not necessarily in identical quantity) to the fixing plant and (second) have a similar maturation date to the fixing plant.[140,141]

Theoretically, the nodulating line in a nonfixing mode is the ideal control since it is identical to the test plant in all respects except in N_2 fixation. However, in soils having indigenous populations of *Rhizobium japonicum* uninoculated controls also nodulate and it may be difficult to ensure 100% nodulation by an applied ineffective *Rhizobium* strain. To avoid these problems many workers[141-145] have used a non-nodulating isoline as the nonfixing control plant. When no non-nodulating isoline of a particular soybean cultivar existed, nonlegumes such as *Hordeum vulgare*[140,146] or *Lolium perenne*[144,147] have been used as the nonfixing control plants. The non-nodulating isoline, even when near isogenic has been shown to differ significantly from the nodulating line in seed dry matter, N yield,[139,148,149] nitrate reductase activity,[150] and fertilizer use efficiency.[142] Rennie et al.[151] were against the use of a non-nodulating isoline as nonfixing control plant since it gave illogical results for %Ndfa (percent plant N derived from the atmosphere) in relation to the increasing rate of applied fertilizer N.

The stable isotope ^{15}N was used to trace the contribution of soil, fertilizer, and atmospheric N to the plant N of soybean (*Glycine max L.* Merrill) by Rennie.[152] It was suggested that the ^{15}N isotope dilution technique was the best technique to quantify N_2 fixation (percent nitrogen derived from the atmosphere), but requires reference to an appropriate nonfixing control plant so that the contribution of soil and/or fertilizer N to the N yield of the fixing plant might be determined. The nodulating soybean line in the nonfixing mode (uninoculated or inoculated with an ineffective strain or *Rhizobium japonicum*) was shown to be the best nonfixing control plant.

Problems associated with accurately quantifying N_2 fixation in the field grown legumes have impeded progress in selecting superior rhizobial strains, legume cultivars and organic conditions that would reduce dependence on fertilizer and soil N. Only two techniques are suitable to estimate N_2 fixation integrated over the growing season, the N balance (NB) method and methods based on the principle of ^{15}N - isotope dilution.[152] The NB method is based on the total NB between N_2-fixing (fs) and nonfixing (nfs) systems thus

$$N_2 \text{ fixed} = \text{yield (fs)} - \text{N yield (nfs)} \tag{1}$$

and

$$\% \text{ Ndfa} = \frac{\text{N yield (fs)} - \text{N yield (nfs)}}{\text{N yield (fs)}} \times 100 \tag{2}$$

where %Ndfa is percent plant N derived from the atmosphere. The NB method, which has been used extensively to estimate N_2-fixation is based on the assumption that the fs and nfs assimilate identical amounts of soil and fertilizer N. Thus, if the N yield of the fs exceeds that of nfs, the difference is attributed to N_2-fixation. However, an increase in total plant N due to rhizobial inoculation may be the result of N_2 fixation or increased efficiency of fertilizer or soil N use. Similarly, equal N yields between inoculated and uninoculated (non-nodulated) plants may indicate either that no N_2 fixation has occurred or fixed N_2 has replaced soil fertilizer N in the nodulated plant. Thus, positive or negative values for the N_2 fixation in the absence of nodulation are sometimes encountered (particularly on high N soils) when N_2 fixation is estimated using the NB method which takes into account residual soil NO_3 available to the fs and nfs has been proposed.[153]

[15]N isotope dilution (ID) has been used to determine the percent plant N-derived from fertilizer (% Ndff), soil (% Ndfs), and atmosphere (% Ndfa) in several annual grain legumes.[151,154] Thus,

$$N_2 \text{ fixed} = \left[1 - \frac{\text{atm\% } ^{15}\text{N ex (fs)}}{\text{atm\% } ^{15}\text{N ex (nfs)}} \right] \times N \text{ yield (fs)} \qquad (3)$$

Where atm% [15]N ex = atm% [15]N excess and

$$\% \text{ Ndfa} = \left[1 - \frac{\text{atm\% } ^{15}\text{N ex (fs)}}{\text{atm\% } ^{15}\text{N ex (nfs)}} \right] \times 100 \qquad (4)$$

The calculation of %Ndfa (Equation 4) is yield dependent, requiring only that the fs and nfs assimilate identical proportions (but not necessarily identical quantities) of N from the soil and/or fertilizer.[155-157] The calculation of N_2-fixed becomes yield dependent because an estimate of N yield of the fs is required (Equation 3). The ID technique have been shown to have an inherently higher precision than yield independent ^{15}N techniques such as the yield-dependent NB method for calculating %Ndfa.[158] Talbott et al.[159] working with six cultivars of soybean (*Glycin max* L Merr) concluded that ID was more precise than NB due to higher spatial variability in soil N uptake than in ^{15}N concentration in non-nodulating soybean line used as nfs.

The one instance when NB and ID would yield identical estimates for %Ndfa and N_2 fixed occurs when the fs and nfs have the identical fertilizer use efficiency (FUE) or when no fertilizer N is applied and when the fs and nfs assimilate identical amounts of soil N. Broadbent et al.[160] incorrectly assumed that the FUE of the fs and nfs were identical. In fact this is seldom the case but Rennie and Rennie[157] showed that if the FUE (nfs) = FUE (fs), then NB would result in a similar estimate of %Ndfa as ID because Equation 3 would reduce to Equation 1.

Many researchers do not have access to ^{15}N or to isotope ratio mass spectrometers and thus rely on the NB method. A comparison of these two techniques of estimating N_2 fixation in several experiments over 3 years showed that NB method generally resulted in a lower estimate of N_2 fixation and was consistently less precise with higher experimental error.[108] The N balance method was more reliable only in experiments when soil N was low, so that nonfixing plants showed signs of N deficiency by anthesis. Negative values for fixing systems or positive values for nonfixing systems for N_2-fixation were found with NB. In one circumstance NB accurately quantify N_2-fixation when the fertilizer use efficiency (or in unfertilized experiments, soil N uptake) of the fixing system was identical to that of nonfixing system.[108] Since this occurs sporadically, NB cannot be used with confidence to estimate N_2 fixation in field grown legumes.

Rennie and Dubetz[106] used the above described methods in order to evaluate the effects of fungicides and herbicides on shoot N, seed yield, nodulation, acetylene reducing activity, and ^{15}N-determined N_2 fixation of soybean. They also emphasized that before studying the affects of pesticides on *Rhizobium*-legume symbiosis prior knowledge of the growth patterns and field conditions of the system should be obtained.[106-108] Prior to analyzing the effects of several pesticides, Rennie and co-workers analyzed the problem of the control plant *Rhizobium* inoculant and the effects on legume as described earlier.[161-164]

VIII. BLUE-GREEN ALGAE

A. Indexes of Growth

The effects of pesticides on blue-green algae have mainly concentrated on two parameters, algal growth and nitrogen fixation. Algal growth can be measured from the dry weight of cells or as packed volume of cells in a volume unit of a given microbial culture. In a growing culture, the dry weight and packed cell volume of cell suspension of the medium increase with time and this increase, termed growth, can be the specific subject of observation. Another characteristic widely used as a measurement of the growth of blue-green algae is an increase in optical density (turbidity) of the suspension of algal cells.

B. Yield as a Growth Indicator

Growth can be expressed as yield or as growth rate. Yield as an expression of organic production is usually given in terms of dry or fresh weight of the organic mass produced over a period of time per unit of volume or unit of area occupied by a given organism. Thus, yield Y can be given by the formula

$$Y = \frac{X_1 - X_0}{A \text{ or } V}$$

where X_o and X_1 are quantitative expression of the mass of cells given usually in terms of dry or fresh weight of cells at the beginning (X_o) and at the end (X_1) of the growth period under observation and A (or V) are correspondingly the area of volume occupied by a given population of microbial cells.

The blue-green algae are cultured in a defined media and treated with pesticides under optimum conditions of growth.[165] Generally in most of the studies total growth and not the yield has been used to measure the effects of pesticides. The serious limitation of yield as an indicator of growth, is that this course of growth, within the period for which yield is estimated, remains unknown. During some portion of that period, growth rate may be lower and net growth may be absent. For this reason if yield has to be taken as parameter for measuring growth, the kinetics of growth of blue-green algae must be studied first.

Measurement of the effect of pesticides on growth of blue-green alga by dry weights of cells involve taking aliquots of an algal suspension from treated as well as controls, drying samples to a constant weight, and expressing the dry weight of cells per unit of volume. Taking of samples which is representative of a given culture, is one of the critical conditions for reliable estimates of dry weight of cells. Adequate stirring of algal suspension and fast pipetting, to prevent cell settling in the process of sampling, are routine requirements for proper sampling. Each measurement of dry weight of cells is usually done on at least three parallel samples (in triplicate). Size of the sample generally depends on the population density of the culture — the aliquot increasing with the decrease in population density.

After taking the aliquot algal suspension, cells are separated from culture medium by centrifugation. The deposit of cells is then resuspended in distilled water and centrifuged again. Washing the cells in distilled water removes salts (present in the nutrient medium) and pesticides (which if adhering to the cells can affect measurement of dry weight). After being washed in distilled water, cells are transferred with a pipette in a small volume of distilled water into a tared weighing dish or aluminum cups. Washing in distilled water must be omitted for marine algae as transferring cells from high salinity medium into distilled water causes plasmolysis resulting in bursting of cells and release of their contents. The weighing dishes may be of the disposable or reusable type. The aluminum, light weight open disposable dishes are greatly used for quick determinations. For precise measurements on small samples, glass covered dishes are preferable. In our laboratory drying at 80°C for

12 hr gave a constant weight. It is essential to avoid excessive temperatures and the extension of drying time beyond that necessary to achieve constant weight, since excessive drying causes changes (oxidation) in the dry weight of cells than the loss of water. Weighing is done after the dish is cooled in a desiccator to room temperature. Dry weight of cells is determined and the overage weight from triplicates is used to calculate dry weight per unit of volume of the original culture.

Instead of centrifugation, cells can be separated from the medium by filteration on a membrane filter and washed on the filter to a constant weight. The dry weight of the algal cells is determined by the substraction of the dry weight of the membrane filter. Among factors most commonly affecting the accuracy of dry weight determinations the following are the most important.

1. Poor sampling resulting in an aliquot of cells which is not representative of the original culture.
2. Loss of the cells during separation of cells from the medium after washing the cells with distilled water.
3. Improper drying, either insufficient drying to bring the sample to constant weight or excessive drying resulting in changes other than removal of water.
4. Improper cooling of the sample before weighing it.

C. Packed Cell Volume

Determination of growth, as increase in packed volume of cells, involves sampling, centrifugation of the sample until deposit of cells is compressed to a constant volume, and calculations of the packed volume of cells per volume or surface unit of algal culture. Sampling does not generally differ from that involved in dry weight measurements. However, depending upon the population density of an algal suspension it may become necessary to condense (or to dilute) the original suspension before using it for packed volume measurements. To avoid changes in the volume of individual cells due to osmotic effects, measurement of packed cell volumes are done in the same nutrient medium in which algae have been grown. Centrifugation to a constant packed volume is done in calibrated capillary tubes. The tubes consist of an enlarged upper portion, the receiver and a lower calibrated portion which is a capillary tube. The capacity of the calibrated section is usually 0.05 mℓ. It is calibrated to 0.001 mℓ and with an approximation, the volume of packed cells can be read to 0.0001 mℓ. A sample of the suspension of algal cells pipetted into the receiver must contain an amount of cells, which when compressed to a constant volume, will occupy not more than 0.05 mℓ (the capacity of the calibrated section). On the other hand, to achieve higher accuracy, the packed volume of cells should be large enough to ocupy at least half of the calibrated section of the packed cell volume tube.

The volume of the algal suspension placed into the receiver, and/or the amount (concentration) of cells in this suspension, can be manipulated by dilution or condensation. Thus, in case of algal suspension of low population density it may become necessary to take a large sample (100 mℓ), to condensate it by centrifugation and to transfer (with a pipette) this condensed suspension into the receiver. Of course, the dilutions (or condensations) must be taken into consideration in final calculations of the packed volume of cells in a given algal suspension. Determinations of packed cell volume is generally done in triplicate and the mean from three measurements is expressed as packed volume of cells cubic centimeter per volume (liter) or surface unit of the original algal culture. The sources of error during packed cell volume determinations are poor sampling, inadequate centrifugation, and delay in reading of packed cell volume causing swelling of cells thus, inflated values of packed cell volume.

D. Optical Density

Of all the indexes of growth, measurements of optical density (turbidity technique) is particularly suitable for determinations of growth rate. The basic advantage of the turbidity technique in growth rate measurements, in addition to its efficiency, is the possibility of taking repeated readings on the increase in turbidity on the same batch of the suspension of microbial cells. However, this is not applicable to blue-green algae, most of which are filamentous.

E. Chlorophyll Measurements as a Parameter of Growth

The chlorophyll content of blue-green algae is a useful index of the biomass. Generally the effects of pesticides on growth are studied by making use of this technique.[166-170] The chlorophyll is extracted with aqueous acetone and the absorbance of the extract is determined with spectrophotometer. When immediate pigment extraction is not possible, the samples may be stored frozen for as long as 30 days if kept in the dark. The ease with which the chlorophylls are removed from the blue-green algae varies considerably with different algae. To achieve complete extraction of the pigments it is usually necessary to disrupt the cells mechanically with a grinder, blender, or sonic disintegrator or by freezing. Grinding is the most rigorous and effective of these methods. The absorbance reading at 665 nm is used for the determination of chlorophyll a.[167] The concentration of the chlorophyll is calculated by inserting the absorbance in the equation or absorbance is directly used for estimating the effects of pesticides.

IX. ANALYTICAL METHODS OF ASSESSING NITROGEN FIXATION

The assessment of the nitrogen fixing ability of a certain system requires a suitable method to directly detect gains in nitrogen or indirectly the nitrogenase activity. The Kjeldahl method for determination of nitrogen content, introduced in 1883, was the only available method until 1940s when the ^{15}N-method was adopted to assess nitrogen fixation in *Azotobacter*. Subsequently different methods were developed to assess ability of an organism to fix nitrogen.

A. Kjeldahl Method

In spite of its limitations the Kjeldahl method is still very useful for assessing the response of pesticides on nitrogen fixation especially in laboratories lacking the specialized equipment required for other methods; although it cannot be used to present conclusive evidence of the capacity of an organism to fix nitrogen. Details on equipment needed, procedure and calculations are presented by Allen[171] and Vincent.[80] The method involves the conversion of nitrogen in biological materials into $(NH_4) SO_4$ by digestion with H_2SO_4 followed by distillation of NH_3 in an alkaline medium. The ammonia is collected in sulfuric acid of known strength (0.05 N) which is back titrated with standard sodium hydroxide solution. While the method is adequate for soils, plant material, and active nitrogen fixers, it is not sensitive enough to measure less than 1 mg of nitrogen.

The biological material about 1 g is placed in a 250 mℓ Kjedahl flask and digestion mixture is added to it. The digestion mixture generally consists of: 20 g $CuSO_4$ 5H$_2$O and 1 g selenium. To one part of this mixture 20 parts of anhydrous Na_2SO_4 or K_2SO_4 are added. Then 5 g of this catalyst mixture is added to the digestion flask along with 5 mℓ of mercuric sulfate solution (concentrated H_2SO_4 and Se accelerate the rate of digestion while methylamines prevent loss of N_2 which may occur if only selinium is present). Also 15 mℓ of concentrated H_2SO_4 and a few glass beads are then added. The digestion flasks are heated, first gently until all the water is removed and charring is completed. The heat is then gradually increased so that the solution is brought to constant boiling with slight bubbling. After complete clearing, boiling is continued gently for another 15 to 20 min. But this time, all of the nitrogen might have been converted into $(NH_4)_2SO_4$, the contents transferred to a 100 mℓ volumetric flask, and the volume made up with distilled water.

The Hoskins steam distillation apparatus is commonly used for distillation. Measured quantity of digested material (10 to 25 mℓ, depending on the N_2 content of the material) is taken in the distillation flask and 10 to 15 mℓ of 40% NaOH added to the sample. The flask containing sulfuric acid and an indicator solution is kept under the condenser of the distillation apparatus. Toshiro's indicator (0.25 g of methylene blue 0.375 g of methyl red, and 300 mℓ of 45% ethanol) or a mixed indicator (0.099 g red bromocresol green and 0.066 g of methyl red in 100 mℓ of ethanol) could be used advantageously. Heating should be carefully regulated to prevent sucking back of the sulfuric acid. After sufficient distillate has been collected, the sulfuric acid is back titrated with standard 0.05 N NaOH (1 mℓ of 0.05 N H_2SO_4 = 0.7 mg ammonium N). The color changes from green to pink or purple depending on the indicator.

Greaves et al.[130] slightly modified the Kjeldahl procedure to study the effects of pesticides on total nitrogen in soil. In this procedure the soil is dried and ground to pass through a 80 to 100 mesh sieve. This soil is placed in 30 mℓ Kjeldahl flask which also receives 1 mℓ of water and 3 mℓ of concentrated H_2SO_4 with a safety pipette. The contents in the Kjeldahl flask are heated gently followed by the addition of half a copper catalyst tablet and half a selenium catalyst tablet. Continuous heating for 4 hr complete the reaction. The contents are cooled with water and transferred into 50 mℓ volumetric flask together with washing and the volume is made up to the mark. Of this solution, 20 mℓ is transferred into the distillation apparatus along with 10 mℓ of 40% NaOH and steam distilled into 5 mℓ of 4% H_3BO_3. The distillate is titrated to pH 5.2 with 0.005 M H_2SO_4 using the automatic titrator.

B. Acetylene Reduction Technique

In 1966 Schollhorn and Burris[172] and Wisconsin and Dilworth[173] discovered in Perth, Australia that the nitrogenase reduced acetylene (C_2H_2) to ethylene (C_2H_4). The method based on this principle was soon developed to measure nitrogenase activity in nonlegume, legume, algae, lake, and soil samples.

Essentially the method involved maintains the test material in a gas-tight container which contains partial pressure of acetylene. At an appropriate time, gas samples may be withdrawn by syringe for immediate or eventual analysis by using flame ionization detector. The amount of C_2H_4 (ethylene) detected is correlated with the intensity of nitrogenase activity in the sample. Values for nitrogenase obtained by this method are conventionally designated as N_2 (C_2H_2) fixing activity. This method has several merits over the Kjeldahl method. This method is also about 10^3 times sensitive than $^{15}N_2$ method and requires simple incubation conditions, the product C_2H_4 can be readily separated from C_2H_2 and the gas samples can be stored for a longer time in gas-tight containers for chromatographic analysis. Further the test material is not sacrificed for analysis and normally can be repeatedly sampled because each analysis requires only a small incubation atmosphere. The disadvantages of the method include the indirect nature of the reaction in the sense that C_2H_2 is not the physiological substate. In most studies the correlation between C_2H_2 reflects nitrogenase activity capable of reducing 1 mol of N_2. This relationship is consistent with the amount of reductant needed since C_2H_2 reduction requires $2e^-$. However, variability in the conversion factor has been observed and the most reliable data for given conditions are obtained by establishing a factor for each particular variation in experimental procedure.

1. Soil Nitrogenase Activity

Obtain a direct soil sample or loosen adhering soil from the root system by gently shaking, washing the root system gently in water and lightly blotting the roots with filter paper.[130]

If the root samples differ greatly in size, similar weights of each should be taken for assay. Soil samples are placed in suitable containers (30 mℓ universal bottles or 60 mℓ wide-mouth reagent bottles depending on sample size) for incubation. The universal bottles are sealed with a metal screw cap with a rubber liner. A small hole is made in the metal cap to expose the rubber liner and allow sampling with hypodermic syringes. Undisturbed samples, such as soil cores or plants growing in pots could be placed in airtight containers of an appropriate size.

Air is withdrawn from the sample container using a hypodermic syringe and replaced with an equal volume of acetylene to give a final concentration of 10% (v/v).

Propane, about 10 ppm, may also be injected. This acts as an internal calibration standard to measure the gas volume of the assay chamber accurately and monitor leaks and other losses of C_2H_2. Acetylene from commercial sources frequently contains impurities and it is preferable to generate the gas as required. Calcium carbide is treated with water and the acetylene produced bubbled through a 10% (w/v) $CuSo_4$ solution followed by an acid dichromate solution. The purified gas is then collected over water.

Chambers containing samples and acetylene are incubated at 20°C for 24 hr and gas samples (1 mℓ) are then withdrawn using hypodermic syringes. These gas samples are analyzed immediately or stored for no more than 3 hr after sealing the needles by stabbing them into rubber bungs or by some arrangements already present in the needles. The ethylene and acetylene contents of the gas samples are analyzed on the gas chromatograph as described earlier. Gas production from samples with no added acetylene should be assayed to account for endogenous production of ethylene. In addition, blank incubation of acetylene alone should be measured.

Gas mixtures can be separated by gas chromatography and then quantified by a flame ionization detector. For acetylene and ethylene various column packing material can be used. A convenient system in use is a 80 to 100 mesh Porapak N or T in a standard stainless steel column at 100°C, with a nitrogen gas flow rate of 25 mℓ/min, using a hydrogen/air flame ionization detector. Most detectors can routinely detect 0.1 ppm C_2H_4 in a 0.5 mℓ gas sample. For porapak N or T, propane has the shortest retention time followed by ethylene and acetylene. As for most concentrations encountered in the assay, peak heights can be taken linearly and related to the concentration with accuracy. Gas chromatographs are coupled with microprocessor where unknown components are analyzed directly both quantitatively and qualitatively.

2. Nonlegume and Legumes

The acetylene reduction assay that the nitrogen-fixing enzyme nitrogenase reduces acetylene has been established for a wide range of biological systems[174-176] and has been extensively used to study the effects of pesticides on asymbiotic nitrogen fixers,[177-185] *Rhizobium*-legume associations,[186-192] and blue green algae.[165] Briefly, free living bacteria, blue-green algae, nodules, soil containing nodules, or whole plant-soil systems are enclosed in gas-tight containers after treatment with pesticide and exposure to an atmosphere containing acetylene. Gas withdrawn from the sample container by syringe and replaced with acetylene to give a final concentration of 10% acetylene may be premixed with the gas phase introduced into the sample container. Samples without acetylene added should be assayed for C_2H_2 produced endogenously. Samples are best incubated at temperatures which are weather constant or related to the in vivo temperature. After a suitable time (30 min) gas samples of 0.5 to 2 mℓ are taken with a syringe. For periods up to 3 hr, the gas sample can be stored by sticking the syringe needle into a rubber bung.

Various problems are associated with nonlegume and legume assay for acetylene reduction. Acetylene and oxygen may at some time interfere with the system. However, Knowles and Denike[193] found no effect of both oxygen and acetylene on nitrogenase activity. Recently

Kalininskaya et al.[194] have described the construction of the incubation vessel for the cultivation of microorganisms which makes it possible to grow them in a specific gaseous phase and to determine the rate of nitrogen fixation using acetylene reduction technique.

Bargersen[195] while working with detached soybean nodules and cultures of *Azotobacter vinelandii* and *Klebsiella aerogenes* concluded that caution was necessary in applying the technique since major errors are likely to result when the conditions in the assay are not carefully matched with the conditions in which nitrogen fixation occurs. Valid use of this method also requires that acetylene does not alter other metabolic activities within the system under investigation and that the ethylene is stable during assay period. However, acetylene has been shown to influence certain microbial activities including nitrogen-fixing bacteria.[196] Proper assay of nitrogen fixation with the acetylene reduction method requires that the acetylene reach the nitrogen fixing sites and that the ethylene formed from the acetylene is insoluble in water.[197] The error associated with failure to consider soluble ethylene becomes greater as the aqueous phase comprises a large fraction of the total volume of the assay chamber. Lee and Watanabe[198] suggested that in laboratory assay of waterlogged soil, the sample should be agitated prior to gas analysis in order to release the trapped ethylene to the gas phase. In the field assay of waterlogged soil system the quality of ethylene remaining in the trapped ethylene to the gas phase can be reduced by stirring the soil prior to gas sampling.[198] The estimation of nitrogen fixation by the acetylene reduction technique is further complicated by the lag in the evolution of ethylene during incubation. Many other drawbacks of the acetylene reduction technique are that it requires more incubation periods and long-term incubation can lead to anomalous results.[174,199-201]

Nodules are generally assayed while attached to the roots. Removing the top has no measurable effects on the short term assay. Some N_2-fixing systems are apparently cold labile and therefore samples should not be stored in cold before the assay. For comparisons, sampling and assays should take place at the same time each day because the nitrogenase activity of nodulated roots may vary with the light intensity (photosynthetic activity) and temperature throughout the day. When possible, nodulated roots should be shaken free from soil and assayed without washing. Activity is diminished when the surface film of water remaining after washing nodulated roots is not carefully removed and drying of nodule also diminishes activity. The activity of nodules is decreased as the partial oxygen pressure (pO_2) falls below 0.01 atm. For rhizosphere, nonsymbiotic organisms, a pO_2 of about 0.04 atm is often optimal. Assay vessels should be large enough to minimize change in pO_2. Liquid samples should be well-shaken during assay. It is preferable to assay soil as intact cores. Samples containing blue-green algae should be illuminated at light intensities corresponding to those *in situ* in the field. For assay of soil-plant systems, soil cores can be placed in airtight plastic containers, such as plastic pots with clear plastic covers connected to the base by a water seal. The containers should be incubated in the light as the activity of rhizosphere-associated nitrogen-fixing bacteria is influenced by photosynthesis. Precautions should be followed when using acetylene.[202] Acetylene may be made by adding water to calcium carbide or is obtainable in commercial cylinders. Commercial acetylene is contaminated with acetone, ethylene, and methane. Further contamination with ethylene may arise from rubber stoppers used as gas exchange ports in the incubation vessels when heated or exposed to trichloroacetic acid.[203] Blank samples must be analyzed to allow co-reaction for ethylene contamination. Acetylene is also highly explosive and caution must be exercized when using it.

Techniques adopted to quantify the amount of N_2 fixed by the many laboratories using this assay vary mainly with material to be assayed, e.g., intact field-grown plants *in situ*,[204] or intact pot-grown plants.[184] Care must be taken when using the acetylene reduction assay to quantify an amount of N_2 fixed per unit area or plant. The theoretical relationship between acetylene reduction and N_2 fixation dictates that 3 mol of C_2H_2 to C_2H_4 and 6 electrons are

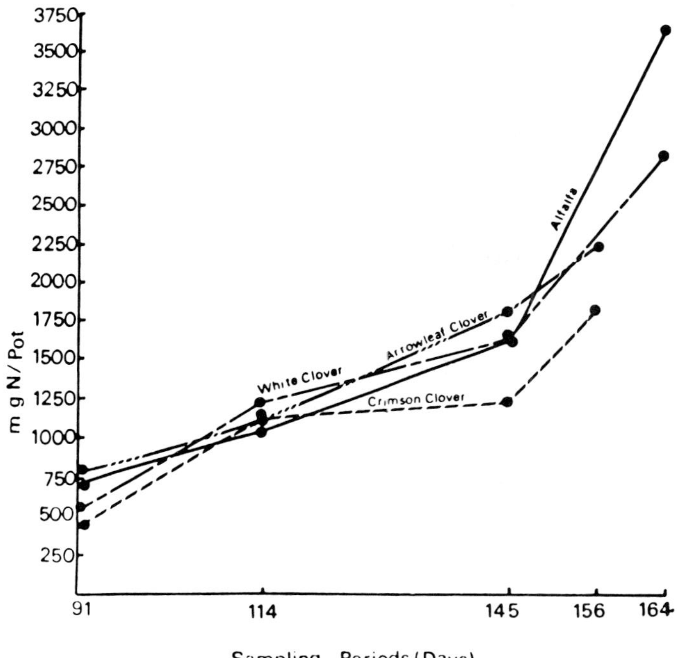

FIGURE 16.   Total N accumulation by alfalfa and clovers grown in pots. (From Giddens, J., University of Georgia Research Bulletin No. 327, Athens, 1—37, 1985. With permission.)

required to reduce N_2 to $2NH_3$. Although some claim good correlations,[176] wide variations in this ratio are found. Hardy et al.[174] reported a conversion ratio of 2.3:1 for soybean. Ham and Caldwell[202] suggested values for soybean ranging from 3.4:1 to 4.4:1. These differences for the same species reflect different assay conditions and perhaps different efficiencies of the strains of *Rhizobium* involved with regard to hydrogen evolution. It is therefore of the utmost importance to establish a quantitative relationship between C_2H_2 reduced and N_2 fixed for the particular legume *Rhizobium* association under study using carefully standardized procedures. Unfortunately, this is done in very few cases.[151] Rates of C_2H_2 reduction are also subject to diurnal variations and should be accounted for when investigating a short-term (0.5 to 1 hr) assay over a full range of 24 hr period.[205]

Giddens[205] has pointed out that cautions must be taken while studying N_2-fixation in legumes by the acetylene reduction technique. The highest N-accumulation rate occurred of the last sampling (mature seed stage) for most of the legumes (Figure 16). The N_2 fixation rate as revealed by acetylene reduction, however, peaked at the 114-day or prebloom stage (Figure 17) indicating that C_2H_2-reduction might not give a good indication of the N_2-fixation rate at late maturity.

3. Blue-Green Algae

Most of the studies on the effects of pesticides on blue-green algae have been carried out under laboratory conditions. In the laboratory algae are grown in a culture medium without combined nitrogen added. In order to measure the effects of pesticides on acetylene reduction the algal suspension (5.0 mℓ) from both control and treated cultures are added to 25 mℓ capacity glass serum bottles fitted with rubber serum stoppers.[206] Then 10% of the air is removed from the serum bottle and replaced by equal volume of acetylene. Activity of nitrogenase is monitored after every 30 min. Ethylene production from acetylene is measured by injection of gas sample taken from sealed flask into gas liquid chromatograph.

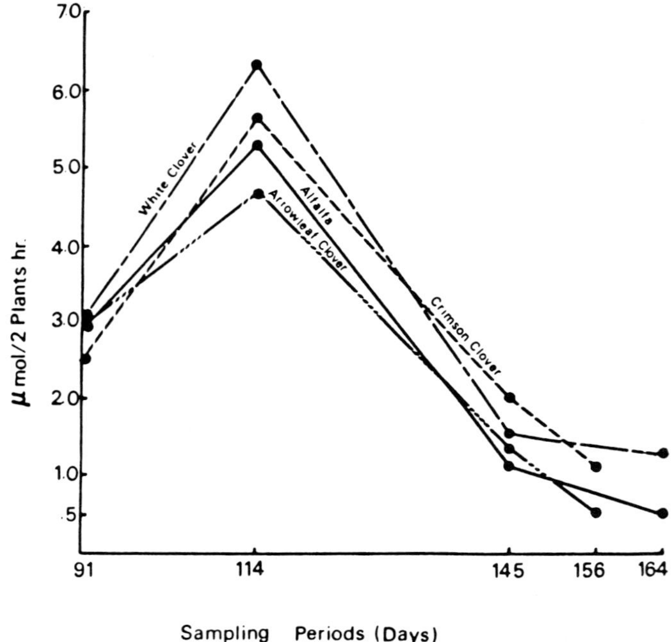

FIGURE 17.   Nitrogen fixation (C_2H_2 reduction) by alfalfa and clovers at different stages of growth. (From Giddens, J., University of Georgia, Research Bulletin No. 327, Athens, 1—37, 1985. With permission.)

C. ^{15}N Technique

The advent of ^{15}N techniques has provided a new approach to many of the problems involved in the study of soil nitrogen. Tracer techniques make possible a reassessment of the principles involved and an extension of experimental facilities. Though the ^{15}N-isotope has not been used extensively to study the effects of pesticides on nitrogen fixation, its importance to confirm the results obtained by other techniques has been recognized by several workers.[199,200]

In biological and especially, agricultural research the most important stable tracer is the heavy nitrogen isotope ^{15}N. There is no other radioactive nuclide of nitrogen that is suitable for work with soils and plants. The half-lives of the existing radioactive nitrogen isotopes are too short (^{13}N, 10 min; ^{16}N, 7 sec; ^{17}N, 4 sec). This means that the stable isotope ^{15}N will remain in use for investigations of events associated with nitrogen cycle.

1. Properties and Analytical Techniques

The stable isotopic tracers are natural constituents of the elements they represent. Thus, ^{15}N is a normal constituent of nitrogen in the atmosphere, soil, and living organisms. Natural nitrogen contains somewhat less than 0.4 atom% ^{15}N. The use of a stable isotope as a tracer requires that it be possible to change the proportions in which the different stable isotopes build up the natural element and that the chemical and biological processes to be studied by the tracer techniques are not able to differentiate between the isotopes. In principle these requirements are contradictory. Both above requirements cannot be fairly fulfilled. If some chemical, physical, or biological processes are able to separate the isotopes, this can to some extent, be expected for all processes of these kinds.[207]

Nitrogen labeling normally means an enrichment of the isotope ^{15}N. The degree of enrichment is usually expressed as the atom percent excess over the natural atom percentage of the isotope in question. Since the ^{15}N percentage in natural nitrogen is 0.4, the atom percent excess of enriched samples will be their atom percentage of ^{15}N minus 0.4 (or more exactly 0.366). So far as its use in tracer work is concerned, the atom percent excess of a stable isotope corresponds to the radiation intensity of a radioactive isotope corrected for background radiation.

Of all the processes used for the enrichment of stable isotopes, thermal diffusion, electromagnetic separation, electrolysis, fractional distillation, and various chemical exchange reactions, the last mentioned group has proved useful with regard to ^{15}N. In the exchange reaction between ammonia gas and the ammonium ion in a solution of ammonium nitrate, ^{15}N enrichment appears in the salt solution. Another example is the exchange reaction between nitric oxide and nitric acid in which the latter component is enriched. By enrichment procedures ^{15}N preparations containing almost pure ^{15}N have been obtained. The cost per unit ^{15}N in excess increases rapidly with the degree of enrichment.

Commercial ^{15}N preparations are not too readily available. They are prepared and sold by a limited number of chemical firms. The number of labeled nitrogen compounds available is also limited, at least in relation to the great number of existing and important nitrogen compounds. However, ^{15}N-enrichment elemental nitrogen as well as most components of commercial nitrogen fertilizers are available. In soil-plant research, the number of enriched nitrogen compounds is satisfactory for most purposes.

An ^{15}N determination in tracer work involves the use of mass spectroscopy. This involves an isotope ratio analysis developed by Rittenberg.[208] Most of the analytical procedures currently employed in ^{15}N work are modifications of methods described in this paper.[209,210] The element to be analyzed in the mass spectrometer is converted into a suitable gaseous form and fed into the sampling tube. There the gas is ionized, converted into a suitable gaseous form, and fed into the electron bombardment. Ions with different masses but the same charge are separated electromagnetically and their relative abundance are determined by measuring the electrical currents they produce when collected on separate electrodes.

In the case of nitrogen the samples to be analyzed are most conveniently converted to elemental nitrogen which is then fed into the spectrometer. The preparation of nitrogen sample for spectrometer analysis is normally performed in two steps, (1) conversion of the labeled nitrogen compounds to ammonium, and (2) conversion of the ammonium to N_2 by oxidation with alkaline sodium hypobromite in the absence of air.

In routine work the analysis required substantial amount of nitrogen (0.5 mg). It is also important that the procedures in handling and converting the labelled nitrogen be practically quantitative and that the N_2 obtained should not contain impurities interfering with the isotope ratio analysis. Considerable care must be taken in every step of the analysis to minimize contamination of the N_2 sample to be analyzed. Severe contaminations are, for example amines, nitrous oxide, and air. Though some other methods may be employed in special cases, the common procedure of converting organic nitrogen to ammonium in soil-plant work is some variant of the Kjeldahl method.[210] The corresponding method of converting oxidized forms of nitrogen-nitrite and nitrate to ammonium is reduction with Devards's alloy in alkaline suspension.

In isotope ratio spectroscopy of nitrogen, the ratio between ions with masses 28 (^{14}N ^{14}N) and 29 (^{14}N ^{15}N) is determined. It is not necessary to measure the ion current corresponding to mass 30 (^{15}N ^{15}N) because equilibrium is supposed to exist in the N_2 gas between ^{14}N ^{14}N and ^{15}N ^{15}N. When the ratio $^{14}N^{14}N/^{14}N^{15}N = R$ is known, the atom percent ^{15}N in the sample can be calculated by the formula $^{15}N = 100/(2R + 1)$. The atom percent excess ^{15}N is then obtained by subtracting the normal abundance of ^{15}N (as determined by the spectrometer employed) from the calculated value of atom percent ^{14}N in the sample.

In the spectrometer analysis there is a permanent risk of contamination by air.[207] The presence of air in the sample can be checked by readings at mass 32 (oxygen) and mass 40 (argon). Depending on the circumstances, a correction for the air present may be undertaken or the samples may be rejected.[207]

In routine work, modern spectrometers are able to detect a change in abundance ratio of 0.2% and to give results that are reproducible with an error less than 0.2% on samples of any isotopic composition down to normal abundance. Since the normal abundance of ^{15}N is 0.4 atom% the limiting excess that can reliably be detected is about 0 to .001 atom% the excess ^{15}N. This means for example, that a nitrogen sample containing 10 atom% excess of ^{15}N may be diluted with natural nitrogen 10,000 times and the presence of labeled nitrogen still be detected in the resulting mixture. It also means that in soil-plant investigations, the sensitivity and accuracy of ^{15}N determinations are clearly better than many of the other sampling and analytical measures included in the experimentation.

For convenience in routine work, samples amounting to 0.5 to 1 mg N have been recommended. However, satisfactory results can be obtained with much smaller quantities of nitrogen. Therefore, extended precautions have to be taken against air leakage and the gas inlet system of the spectrometer may require modifications. Since the method of ^{15}N labeling requires a substantial amount of nitrogen to be introduced into the spectrometer tube, it also requires some special considerations regarding the more-or-less conventional experimental and analytical techniques to be applied in the investigation. In every step of the experimental procedures, sufficient amounts of nitrogen to make possible the final isotope ratio determinations have to be secured. This often means that comparatively large samples have to be used and that time-saving micro-methods (colorimetric) for the determination of nitrate, nitrite, and ammonium nitrogen are of little or no advantage. Too small amounts of prepared soil and too low ^{15}N excesses risk the results of the investigations. However, too large samples and high excesses waste isotope material.

Mass spectrometers are expensive instruments both to acquire and to operate. For the management of the instrument, a skilled and specialized technician is desirable. The number of samples that can be put through per day is limited. Isotope ratio analysis of stable nitrogen though, are much less sensitive and the use of stable isotopes involves no health hazards, no risks of radiation damage to biological materials, and no time limits on the duration of experiments.

It is undoubtedly expensive to use ^{15}N in plant and soil research. Obtaining and operating a spectrometer is the main cost. Compared with the instrumentation costs and the costs of sample preparation, the cost of obtaining the necessary ^{15}N-enriched isotopic material is normally not decisive in research work confine to a laboratory or even to pot experiments. Thus, if analytical facilities are available, one should not hesitate to use the ^{15}N isotope when it can be used to get an advantage in the collection of information.

In most applied research on soil nitrogen, field experiments are needed. Laboratory and pot experiment findings have to be varified and evaluated under field conditions. In field work, applications of ^{15}N techniques are seldom feasible. A normal field experiment requires considerable amounts of labeled nitrogen and the isotope costs are hard to overcome. A possible intermediate between pot and regular field experiments which limit the isotope cost is the use of different types of frame experiments. In these experiments tagged topsoil is kept in small frames embedded in the natural topsoil layer and resting upon the natural subsoil in the field.

2. Principles of Tagging with ^{15}N

The introduction of labeled nitrogen into the soil systems to be studied can be performed in several ways.[211] Often the most convenient way is to start the soil investigations by adding an inorganic nitrogen compound, ammonium salt, or a nitrate. Tagged plant material is readily produced by growing the desired plants in soil-less cultures or in a soil poor in a plant-available nitrogen fertilized with a labeled inorganic nitrogen compound. The tagged plant material may then be added to the test soil in a variety of ways, in its complete conditions or fractionated by biological or chemical means. Defined organic nitrogenous compounds may be isolated from the tagged plant material and used for subsequent soil studies.

Soils in which plants fertilized with labeled nitrogen are grown, automatically become tagged in a natural but complicated way. Some of the added fertilizer nitrogen, may be chemically freed to the mineral or the organic matter of the soil. Tagged plant residues and root exudates, as well as decomposition products of these materials remain in the soil. Microbes usually immobilize some of the fertilizer nitrogen and this nitrogen remains as tagged microbial biomass or decomposition products after the harvest of the crop. Soil tagged in this way may primarily be used to study residual effects of fertilizer nitrogen and can also be used for turnover and humus formation studies. A drawback in many of these studies is the complexity and vagueness of the starting conditions.

Better defined starting conditions can be obtained in two ways. In both cases the soil under study is incubated in the absence of growing plants. The first approach involves the addition of defined-tagged organic nitrogen compounds to the soil. The added material is attacked by soil microflora, partly mineralized, and partly transformed into tagged microbial biomass; also it is possibly transformed into primary nonmineralized residues stabilized against further microbial attack (humus formation).

In the second approach similar results of tagging are attained by adding tagged inorganic nitrogen compounds to the soil together with a non-nitrogenous energy source, thus stimulating immobilization. When a suitable energy source, for example, sugar or cellulose, is readily available to the microflora, the tagged inorganic nitrogen is rapidly and completely transformed into tagged microbial biomass. The tagged soil is then ready for many kinds of nitrogen transformation studies.

The tagging methods indicated are to be considered only as examples of possible tagging procedures, since there are many other ways of solving the tagging problem. The tagging procedure is an important part of ^{15}N investigation.

3. Nitrogen Fixation

The process of nitrogen fixation by which elemental nitrogen is conveyed into biochemical transformations involves a reduction of elemental nitrogen to the ammonium level. The fixation process seems to be of the same general type whether it is performed symbiotically or by free-living organisms. Research into elemental nitrogen fixation has two main purposes, to clarify the mechanisms of the fixation process theoretically and to determine how the process works in the complicated biological system represented by soil. In clarifying the fixation process, tracer nitrogen is of great value in regard to the intermediate stages of the process, for example, the site of the process in legume symbiosis, and so on.

The main aspect, the function of the fixation process in the soil system, is of central interest in studies dealing with soil nitrogen. In principle, tracer work has brilliant possibilities for the elucidation of this function; however, few tracer studies on these problems have been performed so far. Model experiments with forest litter and forest soils have been performed by Huser,[212] whose results indicate a small and slow nonsymbiotic fixation in such systems. An interesting study of the energetic conditions of nonsymbiotic nitrogen fixation in soils was performed by Delwiche and Wijler.[213] In studies of this type,[214] it is necessary to use a gas phase of atmospheric composition containing tagged elemental nitrogen. The experiments must be continued for a considerable length of time. The special problems encountered in the establishment and maintenance of this gas phase of the fixation experiments have hampered the experimental work.

More interest has been devoted to practical applications and details of the nitrogen supply to legume crops.[162-164] By addition of tagged inorganic nitrogen supply to legume cultures, the relations between nitrogen fixation and use of inorganic nitrogen compounds by the inoculated plants have been studied and depressing influence of inorganic nitrogen compounds on the fixation process has been substantiated. An interesting related question has been the study of the competition for inorganic nitrogen compounds between nonlegumes and inoculated legumes, and the nitrogen fixation of the latter in different competitive situations.[215] This has already been discussed in detail.

4. Mineralization and Immobilization

It has been a general hope that fertilizer evaluation work with the use of labeled nitrogen could be simplified and performed without including unfertilized controls in the experimental set up. However, it was soon proved that the two methods of measuring the uptake of fertilizer nitrogen by crops, the conventional (indirect) method of determining the difference between the fertilized treatment and the unfertilized control, and the new (direct) method of determining the amount of labeled fertilizer nitrogen taken up by the crop, seldom produced the same results. The indirect method regularly gives higher figures than the direct method.[211] These results from plant uptake confirmed by laboratory incubation experiments in which the size of the inorganic nitrogen pool (ammonium plus nitrate nitrogen) was followed by two methods.[194,199] The difference appeared to be intriguing and has initiated much of the work on soil hitherto performed. After some time the findings were related to the mineralization-immobilization turnover and taken as an indication of process.[215]

a. Turnover Studies

After the mineralization-immobilization theory was established as an explanation of the phenomenon, one of the main subjects of study using ^{15}N in biochemical soil research was the transformation of nitrogen between inorganic and organic nitrogen components of the soil. Many investigators have stopped with the mere demonstration or even indication of the turnover process.[215] It has been hard to coordinate the nitrogen transformations with other features of biological activity, especially the energy transformations of the heterotrophic soil microflora, its proliferation, dying back, and renewal.

A lasting result of this rather primitive work is that the description of the resulting effects of soil microbial activities in terms of net effects, net mineralization, and net immobilization, has become well established. However, some interesting attempts have been made to extend the experimental aims. It could be inferred that the turnover processes would lead to some kind of equilibrium between the inorganic pool of soil nitrogen and the heterotrophic biomass. Experimental evidence of this was obtained in a model experiment by Jansson et al.[216] Tagged inorganic nitrogen was added to decomposing plant residue in quantities in excess of the needs for optimum rate of decomposition. After regular time intervals of incubation, the organic and inorganic phases tended to approach equivalence as a result of nitrogen turnover. For this to happen, the original plant nitrogen must be largely decomposed (mineralized) by the microflora, the resulting ammonium nitrogen must become a part of the active inorganic pool and the microbes must obtain their nitrogen for cell growth from this active pool. These results were obtained with oat straw in combination with an inorganic pool of ammonium nitrogen.

In soil the conditions are more complicated than in the model decomposition system. For example, nitrification is regularly occurring in normal soils. The relation of nitrate and the process of nitrification, the mineralization-immobilization turnover as revealed by tracer work appears to be an intriguing question.[217] In the tracer work performed by Jansson et al.[216] it was indicated that when a choice was possible the heterotrophic soil microflora markedly preferred ammonium nitrogen. However, when nitrate was available but not ammonium the former was properly utilized. Further in the competition for the mineralization outflow of ammonium nitrogen among heterotrophic microflora, nitrifiers, and higher plants, it was indicated that heterotrophic microflora were successful and principally filled its requirements regardless of the other competitors. In using the left overs, the nitrifiers seemed to be more effective.

In consequences of the preference of heterotrophic flora for ammonium and the competitive relationships mentioned are often somewhat confusing with regard to the evaluation of the mineralization-immobilization turnover. Ammonium nitrogen represents the normal and main pathway of the turnover. Under conditions of net mineralization, nitrification results in a withdrawal of inorganic nitrogen from the turnover process. Plants are the regular consumer of this nitrate nitrogen. The nitrate pool, however, is always readily available to heterotrophic flora. Under conditions of net immobilization, the microflora draw nitrate nitrogen which reappears in the nitrate pool now diluted with nitrogen from other sources that participated in the turnover.[211]

Under continuous net mineralization, which is represented by the prevailing situation in soil, the nitrate pool principally stays off the pathways of the turnover process. A consequence of this is that the difference between the direct and indirect method of fertilizer nitrogen evaluation is less in the case of nitrate fertilizers than in the case of ammonium fertilizers. With regard to the latter form of fertilizer nitrogen, the difference is less in nitrifying soils than in non-nitrifying soils. These conditions have been indicated in a variety of investigations, but they have seldom been considered. However, Broadbent and Tyler have objected to the interpretation given above.[217,218]

b. Priming Action

Results from early mineralization studies starting with readily decomposable tagged organic material suggested that the (energetic) materials added by way of soil microflora increased the mineralization of nitrogen organically bound in soil.[217] These findings prompted further investigations and the phenomenon has been priming action.[218] First, this priming action has been connected with the carbon balance of the soil[219] and related to the nitrogen balance.[220]

It is important to differentiate between priming action and mineralization immobilization turnover, both in the planning and interpretation of experimental work. Especially in the early stages of incubation experiments, the turnover process normally gives results that can be erroneously interpreted as priming action.[211] In some investigations such misinterpretations probably have occurred; they have dealt with an apparent priming action.

Priming action leading to a net decrease in the organic nitrogen content of the soil is not to be expected from any theoretical point of view. Findings of this kind may be attributable to experimental error. However, priming action is a phenomenon that often can be expected and it can be positive as well as negative. All soils contain a native heterotrophic biomass. It is quite natural that this biomass will react with the addition of energetic material. This reaction may include variations in the mineralization of biomass carbon and nitrogen, that gives rise to priming action. For example, a change in the microbial species building up the biomass may occur. The defeated species may die and be liable to mineralization.

c. Effects of Inorganic Nitrogen

Another finding in tracer work on soil nitrogen related to the mineralization-immobilization turnover as well as to the priming action is that increasing additions of tagged inorganic nitrogen compounds regularly cause increments in the nontagged inorganic nitrogen pool of the soil and, consequently, in plant uptake of such nitrogen.[218] This phenomenon looks even more puzzling than the priming action as discussed earlier.

In elaborate investigations Jansson[211] found that the phenomenon was related to microbial activity of the soil. It did not appear in soil samples with no biological activity as revealed by carbon mineralization. It was considered to be a feature of the mineralization-immobilization turnover revealed under special circumstances, namely, in connection with a temporary inorganic nitrogen pool of considerable size. Such a pool is created by fertilizer additions. If the mineralization outflow of nontagged nitrogen meets and becomes equilibrated with a large tagged inorganic pool before the occurrence of immobilization withdrawal, more nontagged nitrogen will accumulate in the pool than if the tagged pool is small and the mineralized nitrogen goes almost directly into the immobilization process. Consequently, this pool dilution effect is related to the size of the pool, that is, to the size of the fertilizer addition. These observations do not indicate that a real stimulation of the mineralization process has occurred. This explanation is principally supported by some other investigators.[215] However, this explanation is not accepted by Broadbent and Tyler[218] who also have studied the phenomenon extensively. According to Broadbent and Tyler the phenomenon is real; mineralization is really stimulated by fertilizer additions. The causes are presumably physical or chemical in nature, for example, salt effects, osmosis, pH changes, and other side effects produced by the nitrogen fertilizers. Such side effects undoubtedly exist but it is improbable that they will appear in such clear-cut and consistent way as revealed by the experimental data. The effects are probably responsible in the few cases in which a decrease in soil inorganic nitrogen has followed the addition of tagged inorganic nitrogen.[220,221] In spite of this criticism, the experimental data of Broadbent and Tyler fits well into the turnover model, including the special behaviour of the nitrate pool.

A communication by Aleksic et al.[222] is worth mentioning in this regard. After having demonstrated the phenomenon in a wide variety of soils, these authors stated that the amount of available inorganic nitrogen in the soil was not necessarily affected by the tagged ammonium sulfate applications. Also, Legg and Sloger[210] considered the phenomenon to be a real increase in the net mineralization of soil nitrogen. They related it to rhizosphere effects. Under conditions of scarce nitrogen supply, the energetic material available for the rhizosphere flora was presumed to be poor in nitrogen and thereby to cause net immobilization. When the nitrogen supply to the plants was ample, no such immobilization occurred and more soil nitrogen was taken up by the plants.

5. Nitrification

The process of nitrification and the biological oxidation of ammonium to nitrate, also involved several problems. The definition itself may in some respect be too narrow, being suited for the conditions in arable soils and limited to the activity of specialized autotrophic bacteria. A more general definition, proposed by Alexander et al.[223] is that nitrification is simply the biological conversion of reduced nitrogenous compounds either organic or inorganic to others having nitrogen in more oxidized state. Several heterotrophs are known to be able to oxidize nitrogen, although the occurrence and importance of this heterotrophic nitrification is still obscure. Another complex of unsolved problems exists regarding the intermediates between ammonium and nitrate in the oxidation process.

In spite of the important role in nitrogen cycling and the obscure points in the process, ^{15}N investigations on nitrification are almost lacking. In addition, some exploratory experiments on the availability of chemically fixed ammonium for the nitrifiers have been performed. In nitrification experiments with tagged ammonium nitrogen added to ammonium-fixing soils, a low availability of this ammonium nitrogen has been found.[211] Jansson[211] found that an established nitrifying flora was more effective in extracting strongly absorbed ammonium than was the chemical extractant employed (1 N KCl solution). This behavior is in contrast to that of heterotrophic soil microflora stimulated by the addition of cereal straw. The chemically extractable ammonium was used up before the nitrification of fixed ammonium began. When nitrification had not been established on extractable ammonium, the nitrifiers seemed not to be able to begin utilizing of the ammonium fixed against the potassium chloride solution.

6. Denitrification

The term denitrification is used here to describe the biological reduction of nitrate and nitrite to volatile gases other than ammonium. The biological process is accomplished by facultative anaerobic organisms capable of using nitrate and nitrite instead of oxygen as hydrogen acceptors. Contrary to the study of nitrification, the study of denitrification has to a considerable degree, include the use of tracer techniques.

With regard to the chemical pathway of the process, it has been established by means of ^{15}N that only a negligible quantity of nitrogen evolved as gas originates from sources other than the nitrate or nitrite.[224] Tracer studies also indicate that nitrous oxide is probably an obligatory precursor of molecular nitrogen in the process.[208] Nitrite oxide detected in denitrification systems in acid media probably originates from a side reaction involving the chemical decomposition acid.[225]

In spite of the general statement that denitrifying organisms are facultative anaerobes, a considerable number of conflicting reports have appeared in which denitrification has been observed under apparently aerobic conditions, some of them on the basis of tracer studies.[226,227] There is probably no contradiction in these findings. Under conditions of high oxygen demand, the transport of oxygen to the sites of the denitrifying organisms may become insufficient and the organisms must utilize nitrite. Such conditions may occur when rapid decomposition is taking place.[227] Special precautions must be taken in experiments on denitrification in which tagged nitrogen gases from the soil are expected to join a confined nitrogen atmosphere around the test soil. The ionic equilibration between the evolved gases and the atmosphere with regard to ^{15}N is a low reaction.

REFERENCES

1. **Bollen, W. B.,** Interactions between pesticides and soil microorganisms, *Ann. Rev. Microbiol.,* 15, 69—92, 1961.
2. **Domsch, K. H.,** Einfliisse von Pflan zenchntzmitteln anf die Bodenmikroflora, M.H. Biol. Bundesanstalt Land. Forstwirtsch., Berlin-Dahlem, Heft, 107, 1963.
3. **Audus, L. J.,** *The Physiology and Biochemistry of Herbicides,* Academic Press, New York, 1964, 163—206.
4. **Alexander, M.,** Microbial degradation and biological effects of pesticides in soil, in *Soil Biology,* Vol. IX, UNESCO, Natural Resources Research, 1969, 209—240.
5. **Helling, C. S., Kearney, P. C., and Alexander, M.,** Behaviour of pesticides in soils, *Adv. Agron.,* 23, 147—240, 1971.
6. **Domsch, K. H. and Paul, W.,** Simulation and experimental analysis of the influence of herbicides on soil nitrification, *Arch. Microbiol.,* 97, 283—301, 1974.
7. **Tu, C. M. and Miles, J. R. W.,** Interaction between insecticides and soil microbes, *Residue Rev.,* 64, 17—65, 1976.
8. **Creaves, M. P., Davies, H. A., Marsh, J. A. P., and Wingfield, G. I.,** Herbicides and soil microorganisms, *CRC Crit. Rev. Microbiol.,* 5, 1—38, 1976.
9. **Wainwright, M.,** A review of the effects of pesticides on microbial activity in soil, *J. Soil Sci.,* 29, 287—298, 1978.
10. **Simon-Sylvestre, G. and Fournier, J. C.,** Effects of pesticides on the soil microflora, *Adv. Agron.,* 31, 1—91, 1979.
11. **Lal, R. and Saxena, D. M.,** Accumulation metabolism and effects of organochlorine insecticides on microorganisms, *Microbiol. Rev.,* 46, 95—127, 1982.
12. **Lal, R.,** Accumulation metabolism and effects of organophosphorus insecticides on microorganisms, *Adv. Appl. Microbiol.,* 28, 149—200, 1982.
13. **Lal, R.,** Factors influencing microbe/insecticide interactions, *CRC Crit. Rev. Microbiol.,* 10, 261—295, 1983.
14. **Lal, R. and Dhanaraj, P. S.,** Cellular aspects of microbe insecticide interactions, *Int. Rev. Cytol.,* 96, 239—262, 1985.

15. **Lal, R.,** *Insecticide Microbiology,* Springer-Verlag, Basel, 1984.

16. **Reddy, B. V. P., Dhanaraj, P. S., and Narayana Rao, V. V. S.,** Effects of insecticides on soil microorganisms, in *Insecticide Microbiology,* Lal, R., Ed., Springer-Verlag, Basel, 1984, 169—201.

17. **Anderson, J. R.,** Effects of pesticides on non-target microorganisms, in *Pesticide Microbiology,* Hill, I. R. and Wright, S. J. Z., Ed., Academic Press, New York, 1978, 313—533.

18. **Greaves, M. P., Cooper, S. L., Davies, H. A., Marsh, J. A. P., and Wingfield, G. I.,** Method of analysis for determining the effects of herbicides on soil microorganisms and their activities, *Agric. Res. Counc. Weed Res. Organ.,* 45, 1—55, 1978.

19. **Domsch, K. H., Jagnow, G., and Anderson, T. H.,** An ecological concept for the assessment of side effects of agrochemical on soil microorganisms, *Residue Rev.,* 86, 66—105, 1983.

20. **Pochon, J.and Chalvignac, M. A.,** Recherches sur la proteolyse bacterienne au sein du soil, *Ann. Inst. Pasteur, Paris,* 82, 690—695, 1952.

21. **Chalvignac, M. A.,** Mesure des pouvoirs anylolytique et proteolytique des terres en aerobiose, *Ann. Inst. Pasteur Paris,* 84, 816—818, 1953.

22. **Lejudie, J. and Chalvignac, M. A.,** Appreciation de l'activite proteolytique de la microflora du soil, *Ann. Inst. Pasteur, Paris,* 90, 359—361, 1956.

23. **Kauffmann, J. and Chalvignac, M. A.,** Recherches sur les methodes de mesure du pouvoir ammonificateur d'une terre, *Ann. Inst. Pasteur Paris,* 79, 228—232, 1950.

24. **Greenwood, D. J. and Lees, H.,** Studies on the decomposition of amino acids in soils, *Plant Soil,* 7, 253—268, 1956.

25. **Verstraete, W.,** Registration requirements for pesticides in relation to soil microorganisms, *Med. Fac. Landbouwwet. Rijksnniv. Gent.,* 45/4, 819—833, 1980.

26. **Walker, N.,** Nitrification and nitrifying bacteria, in *Soil Microbiology,* Walker, N., Ed., Butterworths, London, 1975, 133—146.

27. **Lees, H. and Quastel, J. H.,** Biochemistry of nitrification in soil. I. Kinetics of and the effects of poisons on soil nitrification as studied by a soil perfusion technique, *Biochem. J.,* 40, 803—815, 1946.

28. **Billen, G.,** A method for evaluating nitrifying activity in sediment by dark (^{14}C) bicarbonate incorporation, *Water Res.,* 10, 51-57, 1976.

29. **Samville, M.,** A method for measurement of nitrification roles in water, *Water Res.,* 12, 843—848, 1978.

30. **Hall, G. H.,** Apparent measured rates of nitrification in the hypoliminion of mesotrophic lake, *Appl. Environ. Microbiol.,* 43, 542—547, 1982.

31. **Belser, L. W.,** Bicarbonate uptake by nitrifiers: effects of growth rate, pH, substrate concentration and metabolic inhibitors, *Appl. Environ. Microbiol.,* 48, 110—114, 1984.

32. **Paul, W. and Domsch, K. H.,** Ein methematisches Modell fur den Ni rifikationsprozeb im Boden, *Arch. Mikrobiol.,* 87, 77—92, 1972.

33. **Ardakani, M. S., Rehbock, J. T., and McLaren, D. A.,** Oxidation of nitrite to nitrate in a soil column, *Soil Sci. Soc. Am. Proc.,* 37, 53—56, 1973.

34. **Knowles, G. A., Downing, L., and Barrett, M. J.,** Determination of kinetic constants for nitrifying bacteria in mixed culture, with the aid of an electronic computer, *J. Gen. Microbiol.,* 38, 263—278, 1965.

35. **McLaren, D. A.,** Temporal and vectorial reactions of nitrogen in soil: a review, *Can. J. Soil Sci.,* 50, 97—109, 1970.

36. **McLaren, D. A.,** Kinetics of nitrification in soil: growth of the nitrifiers, *Soil Sci. Soc. Am. Proc.,* 35, 91—95, 1971.

37. **De Barjac, H.,** La puissance denitrificante du soil. Mise au point Disune technique evaluation, *Ann. Inst. Pasteur Paris,* 83—207—212, 1952.

38. **Todd, R. L. and Nuner, J. H.,** Comparison of two techniques for assessing denitrification in terrestrial ecosystem, *Bull. Ecol. Res. Comm. (Stockholm),* 17, 277—278, 1973.

39. **Wheeler, B. E. J.,** Denitrification in percolated soils, *Plant Soil,* 19, 219—232, 1963.

40. **McGarity, J. W.,** Denitrification studies on some South Australian soils, *Plant Soil,* 14, 1—21, 1961.

41. **McGarity, J. W., Gilmour, C. M., and Bollen, W. B.,** Use of an electrolytic respirometer to study denitrification in soil, *Can. J. Microbiol.,* 4, 303—316, 1958.

42. **Wijler, J. and Delwiche, G. C.,** Investigations on the denitrifying process in soil, *Plant Soil,* 5, 155—169, 1954.

43. **Buchanan, R. E. and Gibbons, N. E., Eds.,** *Bergey's Manual of Determinative Bacteriology,* 8th ed., Williams & Wilkins, Baltimore, 1974.

44. **Skinner, Q. D., Adams, J. C., Rechard, P. A., and Bettle, A. A.,** Enumeration of selected bacterial populations in a high mountain watershed, *Can. J. Microbiol.,* 20, 1487—1492, 1974.

45. **Patriquin, D. G. and Knowles, R.,** Denitrifying bacteria in some shallow-water sediments: enumeration and gas production, *Can. J. Microbiol.,* 20, 1037—1041, 1974.

46. **Focht, D. D. and Joseph, H.,** Degradation of 1,1-diphenylethylene by mixed denitrifying bacteria, *Soil Sci. Soc. Proc.,* 37, 698—699, 1973.

47. **Gamble, T. N., Betlach, M. R., and Tiedje, J. M.**, Numerically dominant denitrifying bacteria from world soils, *Appl. Environ. Microbiol.*, 33, 926—939, 1977.
48. **Van Olden, E.**, Menometric investigations on bacterial denitrification, *Proc. Sec. Sci. Kninkl. Akad. Welenschap.*, 43, 635—644, 1947.
49. **Best, A. N. and Payne, W. J.**, Preliminary enzymatic events in asparagine-dependent denitrification by *Pseudomonas perfectomarinus*, *J. Bacteriol.*, 89, 1051—1054, 1965.
50. **Chung, C. W. and Najjar, V. A.**, Cofactor requirements for enzymatic denitrification. II. Nitric oxide reductase, *J. Biol. Chem.*, 218, 627—632, 1956.
51. **Rhodes, M. E., Best, A. N., and Payne, W. J.**, Electron donors and cofactors for denitrification by *Pseudomonas perfectomarinus*, *Can. J.Microbiol.*, 9, 799—807, 1963.
52. **Verhoeven, W.**, Studies on true dissimilatory nitrate reduction. V. Nitrite oxide production and consumption by microorganisms, *Antonie van Leeuwenhoek*, 22, 385—406, 1956.
53. **Hollis, O. L.**, Separation of gaseous mixtures using porous polyaromatic beads, *Anal. Chem.*, 38, 309—316, 1966.
54. **Focht, D. D.**, Methods for analysis of denitrification in soils, in *Nitrogen in the Environment*, Nielson, D. R., and MacDonald, J. G., Eds., Vol. 2, Academic Press, New York, 1978, 433—490.
55. **Garcia, J. L.**, Influence de la rhizosphere du riz sur l'activite denitrifinate protentielle des soils de rizieres du Senegal, *Oecol. Plant*, 8, 315—323, 1973.
56. **Garcia, J. L.**, Sequence des produits formes au cours de la denitrification dans les sols des sols de rizieres du Senegal, *Ann. Microbiol.*, 124 B, 351—362, 1973.
57. **Garcia, J. L.**, Reduction de loxyde nitreuse dans les sols des rizleres du Senegal: Mesure de l'activite denitrifiante, *Soil Biol. Biochem.*, 6, 79—84, 1974.
58. **Garcia, J. L.**, Evaluation de la denitrification dans les rizeres par la methnode de reduction de N_2O, *Soil Biol. Biochem.*, 7, 251—256, 1975.
59. **Garcia, J. L.**, Production d'oxyde nitrique dans les sols de riziere, *Ann. Microbiol.*, 127A, 401—414, 1976.
60. **Garcia, J. L.**, Analyse de different groupes composant la microflore denitriflante de sols de riziere du senegal, *Ann. Microbiol.*, 128A, 433—446, 1977.
61. **Garcia, J.**, Etude de la denitrification chez une bacterie themophile sporulac, *Ann. Microbiol.*, 128, 447—458, 1977.
62. **Georing, J. J. and Dugdale, R. C.**, Denitrification rates in an island bay in the equatorial Pacific Ocean, *Science*, 154, 505—506, 1966.
63. **Georing, J. J. and Dugdale, V. A.**, Estimates of the rates of denitrification in a subarctic lake, *Limnol. Oceanog.*, 11, 113—117, 1966.
64. **Broadbent, F. E. and Clark, F. E.**, Denitrification, in *Nitrogen*, Bartholomew, W. V. and Clark, F. E., Eds., American Society of Agronomy, Madison, Wis., 344, 1965.
65. **Delwiche, C. C. and Bryan, B. A.**, Denitrification, *Ann. Rev. Microbiol.*, 30, 241, 1976.
66. **Focht, D. D., Stolzy, L. H., and Meek, B. D.**, Sequential reduction of nitrate and nitrous oxide under field conditions as brought about by organic amendments and irrigation management, *Soil Biol. Biochem.*, 11, 37—46, 1979.
67. **Rolston, D. E., Fried, M., and Goldhamer, D. A.**, Denitrification measured directly from nitrogen and nitrous oxide gas, *Soil Sci. Soc. Am. J.*, 40, 259—266, 1976.
68. **Rolston, D. E., Hoffman, D. L., and Toy, D. W.**, Field measurement of denitrification. I. Flux of N_2 and N_2O, *Soil Sci. Soc. Am. J.*, 42, 863, 1978.
69. **Birch, H. F.**, The effect of soil drying on humus decomposition and nitrogen availability, *Plant Soil*, 10, 9, 1958.
70. **Rolston, D. E. and Cervelli, S.**, Denitrification as affected by irrigation frequency and applied herbicides, in *Agrochemical Residue Biota Interactions in Soil and Aquatic Ecosystems*, IAEA, Vienna, 1980, 189—199.
71. **Augier, J.**, A propos de la numeration des *Azotobacter* en milieu liquid, *Ann. Inst. Pasteur Paris*, 91, 759—765, 1956.
72. **Timonin, M. I.**, The interaction of higher plants and soil microorganisms. I. Microbial population of rhizosphere of seedlings of certain cultivated plants, *Can. J. Res.*, 18C, 307—317, 1940.
73. **Reyes, A. A. and Mitchell, J. E.**, Growth response of several isolates of *Fusarium* in rhizospheres of host and non host plants, *Phytopathology*, 52, 1196—1200, 1962.
74. **Cook, F. D. and Lochead, A. G.**, Growth factor relationships of soil microorganisms as affected by proximity to the plant root, *Can. J. Microbiol.*, 5, 323—334, 1959.
75. **Stover, R. H. and Waite, B. H.**, An improved method for isolating *Fusarium* spp. from plant tissue, *Phytopathology*, 43, 700—701, 1953.
76. **Singh, K. G.**, Comparison of techniques for the isolation of root-infection fungi, *Nature (London)*, 206, 1169—1170, 1965.

77. **Starkey, R. L.,** Some influences of the development of higher plants upon the microorganisms in the soil. VI. Microscopic examination of the rhizosphere, *Soil Sci.*, 45, 207—249, 1938.

78. **Parkinson, D.,** New Methods for the qualitative and quantitative study of fungi in the rhizosphere, *Pedologie Gand.*, 7, 146—154, 1957.

79. **Jonson, L. F. and Curl, E. A.,** *Method for the Research on the Ecology of Soil Borne Plant Pathogenes,* Burgess Publilshing, Minneapolis, Minn., 1972.

80. **Vincent, J.M.,** *A Manual for the Practical Study of Root-nodule Bacteria,* IBP Handbook No. 15, Blackwell Scientific, Oxford, 1970.

81. **Date, R. A. and Vincent, J. M.,** Determination of the number of root nodule bacteria in the presence of other organisms, *Aust. J. Exp. Agric. Anim. Husb.*, 2, 5—7, 1962.

82. **Davis, R. J.,** Resistance of rhizobia to antimicrobial agents, *J. Bacteriol.*, 84, 187—188, 1962.

83. **Pattison, A. C. and Skinner, F. A.,** The effects of antimicrobial substances in Rhizobium spp. and their use in selection media, *J. Appl. Bacteriol.*, 37, 235—250, 1973.

84. **Graham, P. H.,** Selective medium for growth of *Rhizobium, Appl. Microbiol.*, 17, 769—770, 1969.

85. **Vincent, J. M.,** *Nitrogen Fixation in Legumes,* Academic Press, New York, 1982.

86. **Brockwell, J.,** Accuracy of plant-infection technique for counting populations of *Rhizobium trifolii, Appl. Microbiol.*, 11, 337—383, 1963.

87. **Fisher, R. A. and Yates, F.,** *Statistical Table for Biological Agriculture and Medical Research,* 6th ed., Oliver Boyel, London, 1963.

88. **Vincent, J. M.,** *A Manual for the Practical Study of Root Nodule Bacteria,* IBP Handbook No. 15, Blackwell Scientific, Oxford, 1962, 125—126.

89. **Tuzimura, K. and Watanabe, I.,** The effect of rhizosphere of various plants on the growth of Rhizobium. III. Ecological studies of root nodule bacteria, *Plant Nutr.*, 8, 13—17, 1962.

90. **Thompson, J. A. and Vincent, J. M.,** Methods of detection and estimation of rhizobia in soil, *Plant Soil.*, 26, 72—84, 1967.

91. **Vest, G., Weber, D. F., and Sloger, C.,** *Soybeans: Improvement, Production, and Uses,* Am. Soc. Agron., Monograph 16, Madison, Wis., 1973, 353—390.

92. **Caldwell, B. E. and Vest, G.,** Nodulation interactions between soybean genotypes and serogroups of *Rhizobium japonicum, Crop Sci.*, 8, 680—688, 1968.

93. **Masterson, C. L. and Sherwood, M. T.,** Selection of *Rhizobium trifolii* strains by white and subterranean clovers, *Ir. J. Agric. Res.*, 13, 91—99, 1974.

94. **Russel, P. E. and Jones, D. G.,** Variation in the selection of *Rhizobium trifolii* by varieties of red and white clover, *Soil Biol. Biochem.*, 7, 15—18, 1975.

95. **Brockwell, J., Diatloft, A., Roughley, R. J., and Date, R. A.,** Selection of rhizobia for inoculants, in *Nitrogen Fixation in Legumes,* Vincent, J. M., Ed., Academic Press, New York, 1982, 173—191.

96. **Bohlool, B. B. and Schmidt, E. L.,** A fluorescent antibody technique for determination of growth rates of bacteria in soil in modern methods in the study of microbial ecology, Roswall, T., Ed., *Bull. Ecol. Res. Comm. (Stockholm)*, 17, 336—338, 1973.

97. **Clayet-Marel, J. C. and Crozat, Y.,** Elude Ecologique en immunofluocerene de *Rhizobium japonicum* dens le sol et al rhizozosphere, *Agronomie*, 2, 243—248, 1982.

98. **Reyes, V. G. and Schmidt, E. L.,** Population densities of *Rhizobium japonicum* strain 123 estimated directly in soil and rhizosphere, *Appl. Environ. Microbiol.*, 37, 854—858, 1979.

99. **Reyes, V. G. and Schmidt, E. L.,** Population of *Rhizobium japonicum* associated with the surface of soil-grown roots, *Plants Soil*, 61, 71—80, 1981.

100. **Robert, G. P., Leps, W. T., Silver, L. E., and Brill, W. J.,** Use of two-dimensional polyacrylamide gel electrophoresis to identify and classify *Rhizobium* strains, *Appl. Environ. Microbiol.*, 39, 414—422, 1980.

101. **Moawad, M. A., Ellis, W. R., and Schmidt, E. L.,** Rhizosphere response as a factor in competition among three serogroups of indigenous *Rhizobium japonicum* for nodulation of field grown soybeans, *Appl. Environ. Microbiol.*, 47, 607—612, 1984.

102. **Barber, L. E.,** *Rhizobium meliloti* distribution in the soil following alfalfa inoculation, *Plant Soil*, 64, 363—368, 1982.

103. **Bushby, H. V. A.,** Changes in the numbers of antibiotic-resistant rhizobia in the soil and rhizosphere of field grown *Vigna mungo* Cv., Regir., *Soil Biol. Biochem.*, 13, 241—245, 1981.

104. **Bushby, H. V. A.,** Colonization of rhizosphere and nodulations of two *Vigna* species by rhizobia inoculated onto seed: influence of soil, *Soil Biol. Biochem.*, 16, 635—641, 1984.

105. **Remirez, C. and Alexander, M.,** Evidence suggesting protozoan predation in *Rhizobium* associated with germinating seeds and in the rhizosphere of beans (*Phaseolus vulgaris* L.), *Appl. Environ. Microbiol.*, 40, 492—499, 1980.

106. **Rennie, R. J. and Dubetz, S.,** Effects of fungicides and herbicides on nodulation and N_2 fixation in soybean fields lacking indigenous *Rhizobium japonicum, Agron. J.*, 76, 452—454, 1984.

107. **Rennie, R. J. and Kemp, G. A.,** ^{15}N-determined time course for N_2 fixation in two cultivars of beans (*Phaseolus vulgaris* I), *Agron. J.,* 76, 146—154, 1984.

108. **Rennie, R. J.,** Comparison of N Balance and ^{15}N isotope dilution to quantify N_2 fixation in field grown legumes, *Agron. J.,* 76, 785—790, 1984.

109. **Schmidt, E. L.,** The ecology of the root nodule bacteria, in *Interaction between Pathogenic Soil Microorganisms and Plants,* Dommergues, Y. R. and Krupa, S. V., Eds., Elsvier Scientific, Amsterdam, 1978, 269—304.

110. **Bohlool, B. B. and Schmidt, E. L.,** Immunofluorescent detection of *Rhizobium japonicum* in soils, *Soil Sci.,* 110, 229—236, 1970.

111. **Schmidt, E. L.,** Quantitative autecological study of microorganisms in soil by immunofluorescence, *Soil Sci.,* 188, 141—149, 1974.

112. **Vincent, J. M.,** The root-nodule bacteria of pasture legumes, *Proc. Linn. Soc. N.S.W.,* 79, 4—32, 1954.

113. **Vincent, J. M.,** Root-nodule symbiosis with *Rhizobium,* in *The Biology of Nitrogen Fixation,* Quispel, A., Ed., North Holland, Amsterdam, 1974, 265—341.

114. **Dudman, W. F.,** Immune diffusion analysis of the extracellular soluble antigens of two strains of *Rhizobium meliloti, J. Bacteriol.,* 88, 782—794, 1964.

115. **Hughes, D. Q. and Vincent, J.M.,** Serological studies on root-nodule bacteria. III. Tests of neighbouring strains of the same species, *Proc. Linn. Soc. N.S.W.,* 67, 142—152, 1942.

116. **Purchase, H. F., Vincent, J. M., and Ward, L. M.,** The field distribution of strains of nodule bacteria from species of *Medicago, Aust. J. Agric. Res.,* 2, 262—272, 1951.

117. **Purchase, H. F., Vincent, J. M., and Ward, L. M.,** Serological studies of the root-nodule bacteria. IV. Further analysis of isolates from *Trifolium* and *Medicago, Proc. Linn. Soc. N.S.W.,* 76, 1—6, 1941.

118. **Vincent, J. M.,** Serological studies of the root-nodule bacteria. I. Strains of *Rhizobium meliloti, Proc. Linn. Soc. N.S.W.,* 66, 145—154, 1941.

119. **Gibbins, L. N.,** The preparation of antigens of *Rhizobium meliloti* by ultrasonic disruption: an anomaly, *Can. J. Microbiol.,* 13, 1375—1378, 1967.

120. **Sinha, R. C. and Peterson, E. A.,** Homologous serological analysis of *Rhizobium meliloti* strains by immunodiffusion, *Can. J. Microbiol.,* 26, 1157—1161, 1980.

121. **Wilson, M. H. M., Humphrey, B. A., and Vincent, J. M.,** Loss of agglutinating specificity in stock cultures of *Rhizobium meliloti, Arch. Microbiol.,* 103, 151—154, 1975.

122. **Humphrey, B. A. and Vincent, J. M.,** Specific and shared antigens in strains of *Rhizobium meliloti, Microbios,* 13, 71—76, 1975.

123. **Vincent, J. M.,** Variations in nitrogen-fixing property of *Rhizobium trifolii, Nature (London),* 153, 496—497, 1944.

124. **Beynon, J. L. and Josey, D. P.,** Demonstration of heterogeneity in a natural population of *Rhizobium phaseoli* using variation in intrinsic antibiotic resistance, *J. Gen. Microbiol.,* 118, 437—442, 1980.

125. **Josey, D. P., Beynon, J. L., Johnston, A. W. B., and Beringer, J. E.,** Strain identification in *Rhizobium* using intrinsic antibiotic resistance, *J. Appl. Bacteriol.,* 46, 343—350, 1979.

126. **Danso, S. K. A. and Alexander, M.,** Survival of two strains of *Rhizobium* in soil, *Soil Sci. Am. Proc.,* 38, 86—89, 1974.

127. **Schwinghamer, E. A.,** Effectiveness of *Rhizobium* as modified by mutation for resistance to antibiotics, *Antonie van Leeuwenhoek J. Microbiol. Serol.,* 33, 121—126, 1967.

128. **Kowalski, M., Ham, G. E., Frederick, L. R., and Anderson, I. C.,** Relationship between strains of *Rhizobium japonicum* and their bacteriophages from soil and nodules of field-grown soybean, *Soil Sci.,* 118, 221—228, 1974.

129. **Lesely, S. M.,** A bacteriophage typing system for *Rhizobium meliloti, Can. J. Microbiol.,* 28, 180—189, 1982.

130. **Greaves, M. P., Lockhart, L. A., and Richardson, W. G.,** Measurement of herbicide effects on nitrogen fixation by legumes, Proc. 1978 Br. Corp. Protection Conf., 581—585, 1978.

131. **Fuquay, J. I., Bottomley, P. J., and Jenkins, M. B.,** Complementary method for the differentiation of *Rhizobium meliloti* isolates, *Appl. Environ. Microbiol.,* 47, 663—669, 1984.

132. **Matsudaira, P. T. and Burgess, D. R.,** SDS microslab linear gradient polyacrylamide gel electrophoresis, *Anal. Biochem.,* 87, 386—396, 1978.

133. **Johnen, B. G. and Drew, E. A.,** Ecological effects of pesticides on microorganisms, *Soil Sci.,* 123, 139—324, 1977.

134. **Buckman, H. O. and Brady, N. C.,** *The Nature and Properties of Soils,* Macmillan, New York, 1969, 453—455.

135. **Criswell, J. G. and Hardy, R. W. F.,** Nitrogen fixation in soybeans, measurement techniques and examples of application, in *World Soybean Research,* Hill, L. D., Ed., Interstate Printers, Danville, Ill., 1976, 108—124.

136. **Davidson, F.,** The influence of rhizobial strain and soybean variety on leghemoglobin, hematin, nodule weights, nitrogen fixation, *Diss. Abstr.,* 4077B, 1972—1973, 136.

137. **LaRue, T. A. and Patterson, T. G.,** How much nitrogen do legumes fix?, *Adv. Agron.*, 34, 15—38, 1981.

138. **Bell, F. and Nutman, P. S.,** Experiments on nitrogen fixation by nodulated lucerne, *Plant Soil*, Special vol., 231—264, 1971.

139. **Weber, C. R.,** Nodulating and non-nodulating soybean isolines, I. Agronomic and chemical attributes, *Agron. J.*, 58, 43—46, 1966.

140. **Fried, M. and Boeshart, H.,** An independent measurement of the amount of nitrogen fixed by a legume crop, *Plant Soil*, 43, 707—711, 1975.

141. **Fried, M. and Middleboe, V.,** Measurement of amount of nitrogen fixed by a legume crop, *Plant Soil*, 47, 713—715, 1977.

142. **Deibert, E. J., Bueriego, M., and Olson, R. A.,** Utilization of ^{15}N fertilizer by nodulating and non-nodulating soybean isolines, *Agron. J.*, 71, 717—723, 1979.

143. **Ruschel, A. P., Vose, P. B., Victoria, R. L., and Salati, E.,** Comparison of isotope techniques and non-nodulating isoline to study the effects of ammonium fertilization on dinitrogen fixation in soybean, *Glycine max*, *Plant Soil*, 53, 513—525, 1979.

144. **Kohl, D. H. and Shearer, G.,** Isotopic fractionation associated with symbiotic N_2 fixation and uptake of NO_3 by plants, *Plant Physiol.*, 66, 61—65, 1980.

145. **Kohl, D. H., Shearer, G., and Harper, J. E.,** Estimates of N_2 fixation based on differences in the natural abundance of ^{15}N in nodulating and non-nodulating isolines of soybeans, *Plant Physiol.*, 66, 61—65, 1980.

146. **Rennie, D. A., Paul, E. A., and Johns, L. E.,** Natural nitrogen-15 abundance of soil and plant samples, *Can. J. Soil Sci.*, 56, 43—50, 1976.

147. **Domenach, A. M., Chalamet, A., and Pachiudi, E.,** Estimation de la fixation d'azote par le sojaa'l aide de deuse methodes d'analysis isotopiques, *C.R. Seances Acad. Sci.*, 289, 291—293, 1979.

148. **Bhangoo, M. S. and Albritton, D. J.,** Nodulating and non-nodulating soybeans isolines' response to applied nitrogen, *Agron. J.*, 68, 642—645, 1976.

149. **Jeppson, R. J., Johnson, R. R., and Hadley, H. H.,** Variation in mobilization of plant nitrogen to the grain in nodulating and soybean genotypes, *Crop Sci.*, 18, 1058—1062, 1978.

150. **Liu, M. C. and Hadley, H. H.,** Relationships of nitrate reductase activity of protein content in related nodulated and non-nodulating soybeans, *Crop. Sci.*, 11, 467—471, 1971.

151. **Rennie, R. J., Rennie, D. A., and Fried, M.,** Concepts of ^{15}N usage in dinitrogen fixation studies, in *Isotopes in Biological Dinitrogen Fixation*, International Atomic Energy Agency, Vienna, 1978, 107—133.

152. **Rennie, R. J.,** Quantifying dinitrogen (N_2) fixation in soybeans by ^{15}N isotope dilution: the question of the nonfixing control Plant, *Can. J. Bot.*, 60, 856—861, 1982.

153. **Stulphagel, R.,** Schatzung der von Ackerbohnen symbiotisch fixierten Sticktoffimenge im Feldversuch mit der erweitenten Differenzmethode, *Z. Acker-Pflanzenb.*, 151, 446—458, 1982.

154. **Vose, P. B., Ruschel, A. P., Victoria, R. L., Saito, S. M., and Matsui, E.,** 15-nitrogen as a tool in biological nitrogen fixation research, in *Biological Nitrogen Fixation for Tropical Agriculture*, Graham, P. H. and Harris, S. E., Eds., CIAT, Cal., 1981, 575—592.

155. **Rennie, R. J. and Kemp, G. A.,** N_2 fixation in field beans quantified by ^{15}N isotope dilution. I. Effects of strains of *Rhizobium phaseoli*, *Agron. J.*, 75, 640—644, 1983.

156. **Rennie, R. J. and Kemp, G. A.,** N_2 fixation in field beans quantified by ^{15}N isotope dilution. II. Effect of cultivars of beans, *Agron. J.*, 75, 645—649, 1983.

157. **Rennie, R. J. and Rennie, D. A.,** Techniques for quantifying N_2 fixation in association with non-legumes under field and green house conditions, *Can. J. Microbiol.*, 29, 1022—1035, 1983.

158. **Rennie, R. J.,** A comparison of ^{15}N-added methods for determining symbiotic dinitrogen fixation, *Rev. Ecol. Biol. Sol.*, 16, 455—463, 1979.

159. **Talbott, H. J., Kenworth, W. J., and Legg, J. C.,** Field comparison of the nitrogen-15 and difference methods of measuring nitrogen fixation, *Agron. J.*, 74, 799—804, 1982.

160. **Broadbent, F. E., Nakashima, T., and Chang, G. Y.,** Estimation of nitrogen fixation by isotope dilution in field and greenhouse experiments, *Agron. J.*, 74, 625—628, 1982.

161. **Rennie, R. J.,** Free nitrogen, in *Nukrop News*, Pulse Growers Assoc., Alberta, 1982.

162. **Rennie, R. J.,** ^{15}N isotope dilution as a measure of dinitrogen fixation of *Azospirillum brasilense* associated with maize, *Can. J. Bot.*, 50, 21—24, 1980.

163. **Rennie, R. J. and Dubetz, S.,** Multistrain vs. single strain *Rhizobium japonicum* inoculants for early maturing (00 and 000) soybean cultivars. N_2 fixation quantified by ^{15}N isotope dilution, *Agron. J.*, 76, 498—502, 1984.

164. **Rennie, R. J., Dubetz, S., Bol, J. B., and Muendel, H. H.,** Dinitrogen fixation measured by ^{15}N isotope dilution in two Canadian soybean cultivars, *Agron. J.*, 74, 725—730, 1982.

165. **Lal, S.,** Interaction of Blue Green Algae, a Ciliate Protozoan and Yeast with DDT, Fenitrothion and Chlorpyrifos, Ph.D. thesis, University of Delhi, India, 1985.

166. **Stratton, G. W., Burrel, R. E., Kurp, M. L., and Corke, C. T.**, Interaction between the solvent acetone and the pyrethroid insecticide permethrin of activities of blue green algae *Anabaena, Bull. Environ. Contam. Toxicol.*, 24, 562—569, 1980.

167. **Stratton, G. W. and Corke, C. T.**, Effect of acetone on the toxicity of atrazine towards photosynthesis in blue green algae, *J. Environ. Sci. Health*, 16B, 21—33, 1981.

168. **Stratton, G. W.**, Interaction effects of permethrin and atrazine combinations towards several nontarget microorganisms, *Bull. Environ. Contam. Toxicol.*, 31, 297—303, 1983.

169. **Stratton, G. W. and Corke, C. T.**, Toxicity of the insecticide permethrin and some degradation products towards algae and cyanobacteria, *Environ. Pollut.*, 19, 71—80, 1982.

170. **Stratton, G. W.**, Effects of herbicide atrazine and its degradation products alone and in combination on photoautotrophic microorganisms, *Arch. Environ. Contam. Toxicol.*, 13, 35—42, 1984.

171. **Allen, M. B.**, Photosynthetic nitrogen-fixations by blue-green algae, *Sci. Monthly*, 83, 100—106, 1956.

172. **Schollhorn, R. and Burris, R. H.**, Study of intermediates in nitrogen fixation, *Fed. Proc.*, 25, 710, 1966.

173. **Dilworth, W. J.**, Acetylene reduction by nitrogen-fixation preparations from *Clostridium pasteurinum, Biochem. Biophys. Acta*, 127, 285—294, 1966.

174. **Hardy, R. W. F., Burns, R. C., and Holsten, R. D.**, Application of the acetylene-ethylene assay for measurement of nitrogen fixation, *Soil Biol. Biochem.*, 5, 47—81, 1973.

175. **Hardy, R. W. F., Holsten, R. D., Jackson, E. K., and Burns, R. C.**, Acetylene-ethylene assay for nitrogen fixation: laboratory and field evaluation, *Plant Physiol.*, 43, 1185—1207, 1968.

176. **Hardy, R. W. F., Burns, R. C., Hebert, R. R., Holsten, R. D., and Jackson, E. K.**, Biological nitrogen fixation; a key to world protein, *Plant Soil*, Special vol., 561—590, 1971.

177. **Govindaraju, K., Vlassak, K., and Heremans, K. A. H.**, Effect of the thiocarbamates, cycloate and diallate and the dithiocarbamates and maneb and thiram on biological nitrogen fixation and nitrogen mineralization in soils, *Med. Fac. Landbowwet, Rijksuniv. Gent.*, 40, 1209—1219, 1975.

178. **Hegazi, N., Monib, M., Balal, M., Amer, H., and Farag, R. S.**, The effects of some pesticides on asymbiotic N_2 fixation in Egyptian soil, *Arch. Environ. Contam. Toxicol.*, 8, 629—635, 1979.

179. **Malanchuk, J. L.and Joyce, K.**, Effect of 2,4-D on nitrogen fixation and carbon dioxide evolution in a soil microcosm, *Water Air Soil Pollut.*, 29, 181—189, 1983.

180. **Tam, T. Y. and Trevors, J. T.**, Effect of pentachlorophenol on asymbiotic nitrogen fixation, *Water Air Soil Pollut.*, 16, 409—414, 1981.

181. **Tam, T. Y. and Trevors, J. T.**, Toxicity of pentachlorophenol to *Azotobacter vinelandii, Bull. Environ. Contam. Toxicol.*, 27, 230—234, 1981.

182. **Tu, C. M.**, Effects of pesticides on acetylene reduction and microorganisms in a sand loam, *Soil Biol. Biochem.*, 19, 451—456, 1978.

183. **Tu, C. M.**, Effects of pesticides on activities of enzymes and microorganisms in a clay soil, *J. Environ. Sci. Health*, 16B, 179—191, 1981.

184. **Vlassak, K. and Livens, J.**, Effects of some pesticides on nitrogen transformations in soil, *Sci. Total Environ.*, 3, 363—372, 1975.

185. **Vlassak, K.**, Effects of some pesticides on biological nitrogen fixation and on mineralization of nitrogen in soil, *Rucz. Glebozn.*, 26, 193—199, 1975.

186. **Fisher, D. J. and Hayes, A. L.**, Effects of some fungicides used against cereal pathogens on the growth of *Rhizobium trifolii* and its capacity to fix nitrogen in white clover, *Ann. Appl. Biol.*, 98, 101—107, 1981.

187. **Faizah, H. W., Broughton, W. J., and John, C. K.**, Rhizobia in tropical legumes-XI. Survival in the seed environment, *Soil Biol. Biochem.*, 12, 219—227, 1980.

188. **Gillberg, B. O.**, On the effects of some pesticides on *Rhizobium* and isolation of pesticide-resistant mutants, *Arch. Mikrobiol.*, 75, 203—208, 1971.

189. **Jensen, H. L.**, The effect of various herbicides on root nodule bacteria, *Tidsskr. Piavl*, 73, 309—317, 1969.

190. **Odeyemi, O. and Alexander, M.**, Use of fungicide resistant rhizobia for legume inoculation, *Soil Biol. Biochem.*, 9, 247—251, 1977.

191. **Malik, M. A. B.and Tesfai, K.**, Compatibility of *Rhizobium japonicum* with commercial pesticide in vitro, *Bull. Environ. Contam. Toxicol.*, 31, 432—437, 1983.

192. **Malik, M. A. B. and Tesfai, K.**, Pesticidal effect on soybean-rhizobia symbiosis, *Plant Soil*, 85, 33—41, 1985.

193. **Knowles, R. and Denike, D.**, Effect of ammonium, nitrite and nitrate-nitrogen on anaerobic nitrogenase activity in soil, *Soil Biol. Biochem.*, 6, 353—358, 1974.

194. **Kalininskaya, T. A., Red'kina, T. V., Belov, Yu. M., Ippolitov, L. T., and Kokunov, A. V.**, Use of acetylene method for enumerating various groups of nitrogen-fixing bacteria by the limiting dilution method, *Mikrobiologiya*, 50, 924—927, 1981.

195. **Bergersen, F. J.**, Physiology of legume symbiosis, in *Interaction between Non-Pathogenic Soil Microorganisms and Plants*, Dommergues, W. R. and Krupa, S. V., Eds., Elsevier, Amsterdam, 1978, 305—33.

196. **Brouzes, R., Mayfield, C. I., and Knowles, R.,** Effect of oxygen partial pressure on nitrogen fixation and acetylene reduction in sandy loam soil amended with glucose, *Plant Soil,* Special vol., 481—491, 1971.

197. **Flett, R. J., Hamilton, R. D., and Campbell, N. E. R.,** Aquatic acetylene-reduction technique: solution to several problems, *Can. J. Microbiol.,* 22, 43—51, 1976.

198. **Lee, K. K. and Watanabe, I.,** Problem of acetylene reduction technique applied to water saturated paddy soils, *Appl. Environ. Microbiol.,* 34, 654—660, 1977.

199. **Nayak, D. N. and Rajaramamohan Rao, V.,** Pesticides and hetero-trophic nitrogen fixation in paddy soil, *Soil Biol. Biochem.,* 12, 1—4, 1980.

200. **Nayak, D. N. and Rajaramamohan Rao, V.,** Pesticides and heterotrophic nitrogen fixation in paddy soil, *Soil Biol. Biochem.,* 14, 207—210, 1982.

201. **Burris, R. H.,** Nitrogen fixation-assay methods and techniques, *Methods Enzymol.,* 24B, 415—431, 1972.

202. **Ham, G. E. and Caldwell, A. C.,** Fertilizer placement effects on soybean yield, N_2 fixation and ^{33}P uptake, *Agron. J.,* 70, 779—783, 1978.

203. **Sprent, J. E.,** *The Biology of Nitrogen Fixing Organisms,* McGraw-Hill, New York, 1979.

204. **Harridge, D. F.,** Assessment of nitrogen fixation, in *Nitrogen Fixation in Legumes,* Vincent, J. M., Ed., Academic Press, New York, 1982, 123—126.

205. **Giddens, J.,** Nitrogen cycling in Georgia Soils, University of Georgia Bulletin No., 327, Athens, Ga., 1—37, 1985.

206. **Stewart, W. D. P., Fitzgerald, G. P., and Burris, R. H.,** Acetylene reduction by nitrogen fixing blue green algae, *Arch. Mikrobiol.,* 62, 336—348, 1968.

207. **Bremner, J. M.,** Total nitrogen, *Agronomy,* 9, 1149—1178, 1965.

208. **Rittenberg, D.,** Preparation and measurement of isotopic tracers, *Ann. Arbor,* 1946.

209. **Smith, J. H., Legg, J. O., and Carter, J. N.,** Equipment and procedure for ^{15}N analysis of soil and plant material with mass spectrometer, *Soil Sci.,* 96, 313—318, 1963.

210. **Legg, J. C. and Sloger, C.,** A tracer method for determining symbiotic nitrogen fixation in field studies, *Proc. Int. Conf. Stable Isot.,* 661—666, 1975.

211. **Jansson, S. L.,** Tracer studies on nitrogen transformations in soil with special attention on mineralization-immobilization relationship, *Am. R. Agr. Col. Sweden,* 24, 101—361, 1958.

212. **Huser, R.,** Zur, mikrobiologischen luftst icktoffbindung in buchenstren und buchenmull, *Plant Soil,* 23, 236—246, 1965.

213. **Delwiche, C. C. and Wijler, J.,** Nonsymbiotic nitrogen fixation in soil, *Plant Soil,* 7, 113—128, 1956.

214. **Chang, P. C. and Knowles, R.,** Nonsymbiotic nitrogen fixation in some quebee soils, *Can. J. Microbial.,* 11, 29—38, 1965.

215. **Walker, T. W., Adams, A. F. R., and Orchiston, H. D.,** Fate of labelled nitrate and ammonium nitrogen when applied to grass and clover grown separately and together, *Soil Sci.,* 81, 339—351, 1956.

216. **Jansson, S. L., Hallson, M. J., and Bartholomew, W. W.,** Preferential utilization of ammonium over nitrate by microorganisms in the decomposition of oat straw, *Plant Soil,* 6, 382—390, 1955.

217. **Broadbent, F. E. and Tyler, K. B.,** Laboratory and greenhouse investigation of nitrogen immobilization, *Soil Sci. Soc. Am. Proc.,* 26, 459—462, 1962.

218. **Broadbent, F. E. and Tyler, K. B.,** Effect of pH on nitrogen immobilization in two California soils, *Plant Soil,* 23, 314—322, 1965.

219. **Jenkinson, D. S.,** The Use of the isotopes in soil organic matter studies, *FAO/IAEA, Vienna, 1966.*

220. **Mortensen, J. L.,** Decomposition of organic matter and mineralization of nitrogen Brookston silt loam and alfalfa green house manure, *Plant Soil,* 19, 374—384, 1963.

221. **Megusar, F.,** Isotope and radiation in soil organic matter studies, FAO/IAEA, Vienna, 1968.

222. **Aleksic, Z., Broeshart, H., and Middleboe, V.,** The effect of nitrogen fertilization on the release of soil nitrogen, *Plant Soil,* 29, 474—478, 1968.

223. **Alexander, M., Marshall, K. C., and Hirsch, P.,** Autotrophy and heterotrophy in denitrification, Trans. 7th Int. Soil Sci. Madison, Wis., 2, 586—591, 1960.

224. **Payne, W. J.,** *Denitrification,* John Wiley & Sons, New York, 1981.

225. **Cady, F. B., and Bartholomew, W. V.,** Sequential products of anaerobic denitrification on Norfolk soil material, *Soil Sci. Soc. Am. Proc.,* 24, 477—482, 1960.

226. **Marshall, R. O., Dishburger, H. J., MacVicar, R., and Hallmark, C. D.,** Studies on the effect of aeration on nitrate reduction by *Pseudomonas* species using ^{15}N, *J. Bacteriol.,* 66, 254—258, 1953.

227. **Allison, M.,** The enigma of the soil nitrogen balance sheets, *Adv. Agron.,* 7, 213—250, 1955.

INDEX

A

G

H

I

J

Jackbean, 35
Japan, blue-green algae in, 51—52, 55

K

Kangaroo-horn, 35
Kidney bean, 35
Kingella, 9
Kjeldahl technique, 13, 109—110
Klebsiella, 14, 15, 18
  aerogenes, 112
  pneumoniae, 15, 20
Korea, blue-green algae in, 51
Korean lespedeza, 35
Kudzu, 35
Kura clover, 36

L

Ladino clover, 34
Lappa clover, 34
Large hop clover, 34
Lathyrus, 21, 33
  hirsutus, 35
  odoratus, 35
  sylvestris, 35
  tingitanus, 35
Lead plant, 36
Lectins
  cross-bridging model and, 38
  Rhizobium-legume association and, 30
Lee cultivar soybeans, rhizobial inoculation of, 45
Lees-Quastel apparatus, 80
  enumeration of denitrifiers with, 89
  enumeration of nitrifiers with, 81
Leghemoglobin, temperature and, 47
Leghemoglobin content, nitrogen fixation measurement and, 104
Legume(s), see also specific legume
  cross-fertilized, 39—40
  distribution of, 39—40
  environmental influences on, 43—47
  growth of, assessment of, 100—103
  nitrogen fixation in, 40—43
    acetylene reduction technique and, 111—113
    measurement of, 103—106
  nodulation of, assessment of, 100—103
  population structure of, 39—40
  Rhizobium and, 14, 21
    evaluation of association of, 94—106
    specificity of association of, 30—31
  self-fertilized, 39
  yield of, assessment of, 100—103
Lens, 33
  culinaris (esculenta), 35
Lentil, 35, 47

Lespedeza
  cuneata, 35
  stipulacea, 35
  striata, 35
  virginica, 35
Lichen phycobionts, 52
Lichens, 14, 56
Lima bean, 35
Linuron, 87, 92
Liquid inoculations, commercial, 38
Lithotrophs, 8
Little bur-clover, 34
Lolium perenne, 105
Lotus
  corniculatus, 36
  uliginosus, 36
Lotus rhizobia, 21
Lupine cross-inoculation group, 35
Lupinus, 21
  albus, 35
  angustifolius, 35
  luteus, 35
  perennis, 35
  polyphyllus, 35
  subcarnosus, 35
Lyngbya 6409, 49

M

Macrozamia, 57
Maize, 47
Malaysia, blue-green algae in, 51
Manometry, 89
Marine forms, heterocystous, 51
Mass, nitrogen fixation measurement and, 104
Mastigocladus, 52
  laminosus, 49, 51
Mat bean, 35
Medicago, 21, 33
  arabica, 34
  denticulata, 34
  falcata, 30, 34, 39
  lupulina, 34
  minima, 34
  orbicularis, 34
  rigidula, 34
  sativa, 34, 39
  scutellata, 34
  tuberculata, 34
Meliatus, 21
Melilotus
  alba, 34
  alba annua, 34
  indica, 34
  officinalis, 34
Methabenzthiazuron, 86—88
Methane oxidizers, 14
Methylosinus microorganisms, 16
Metobromuron, 86
Microaerophilic bacteria, nitrogen-fixing, 14